Advance Praise for John Ross and *The Storeroom Roadmap*

John and I met a several years ago as he was auditing a third-party storeroom service program that was part of an MRO service agreement I was managing for a large manufacturing client. From this brief encounter, John and I have shared countless phone calls and emails to discuss strategies on how to improve the functions of the maintenance storeroom.

I have spent just over 26 years running, optimizing, and developing storeroom management services. John's depth of knowledge on this topic is unparalleled. His understanding of the criticality of the maintenance storeroom's processes and operations is an essential guide in the journey toward asset optimization and provides a foundational path to storeroom greatness.

In the world of storerooms, there is an entire ecosystem (integrated supply) focused almost exclusively on the MRO supply chain impact to an organization's bottom-line. However, the link between the storeroom and maintenance organization is often overlooked as these services are often primarily focused on the high-volume, low-cost materials that make up a large share the transactional workload in a business. Many of the third-party storeroom service providers manage the parts and transactions in exchange for exclusive rights to the MRO purchases, and the deal is negotiated and often awarded to the "lowest cost" bidder. While the services typically deliver an improvement over the client's current state regarding supply chain, transactional workload, and storeroom functionality, the results may not be focused on the benefits that are most important to the maintenance organization. This book will provide the reader the correct strategy, processes, and metrics, whether you self-manage or outsource your storeroom management program.

—**Will Brummett**, Vice President, Business Development,
North America, RS Integrated Supply

John and I first met in early 2015. John mentioned that he was designing curriculum for maintenance storeroom training. One of the projects that I had overseen in the oil production facility that I had previously managed was the construction of a new maintenance and operations spares complex. Of course, our common interest led to many conversations about the value of the MRO Storeroom. John's passion and real-world experiences bring the concepts, processes, and practices of a world-class storeroom to ground level. I can personally attest to John's knowledge and practical application captured in these pages. If you want a highly effective and efficient storeroom, this book will provide the roadmap to get you there.

—**Gary Rigdon**, CMRP

I have had the privilege and pleasure of working with Dr. John Ross for close to a decade. It's truly a privilege to work with someone with John's background and knowledge in all aspects of maintenance and reliability. John is not only one of the world's preeminent experts in maintenance and reliability and a lifelong student, but also a practitioner. Having implemented the practices he teaches, John "knows where of he speaks."

Some of the pleasure of working with John comes from his unique style of teaching. I have been a member of captive audiences at many of John's workshops and seminars. John presents material in a conversational style that incorporates humor and anecdotes. Known to wax philosophical at times, John is as likely to quote Dr. Emmet Brown as he is to quote Thoreau. John appears equally at home with shop floor personnel and planners/schedulers as he is with maintenance managers and C-suite executives.

After years of maintenance initiatives and training seminars, our maintenance department continued to struggle to make the advancements we were seeking. After reading John's first book, *The Reliability Excellence Workbook: From Ideas to Action*, I contacted John. Working with facility and maintenance managers from our five facilities, John helped us develop a strategic plan to move toward best practices in maintenance and reliability. The plan we developed included input from all our stakeholders and is specific to our unique environment. Maintenance is a business, and, with few exceptions, successful businesses have documented detailed processes. With John's help, we developed detailed processes for all the primary maintenance functions, all tailored to our specific environment.

During our company's journey with John, my greatest pleasure has been what John refers to as our "windshield time," the time we spend together traveling between our facilities. We have had many hours of this treasured time, and the conversation never lags. To have the opportunity to share John's passion for maintenance and reliability is truly a pleasure and a privilege. John's new book, *The Storeroom Roadmap: 32 Processes for Efficiency and Sustainability*, is a must read.

—**Doug Posey**, Packaging/Warehouse Maintenance Superintendent, American Crystal Sugar Company

As a business owner, cofounder, and expert in the maintenance and reliability sector, I cannot wait to see others dive into the captivating world of maintenance and reliability, where each machine is a vital piece of the puzzle, and effective inventory management is the key to success. Discover how the right part at the right time can make all the difference in this fascinating journey where foresight meets preparedness.

I am thrilled to recommend *The Storeroom Roadmap* to you and your team. The author, who happens to be a long-time friend and maintenance and reliability professional, expertly unravels the complexities of maintaining a well-organized storeroom operation, offering insights on transforming challenges into opportunities for efficiency and success. It would be a challenge to find someone more passionate and knowledgeable about storeroom best practices than Dr. John Ross. Prepare to revolutionize your approach to inventory management and lay the foundation for a more resilient and reliable future.

—**Justin Smith**, President/CEO, TBR Strategies, LLC

The

STOREROOM ROADMAP

32 PROCESSES FOR EFFICIENCY AND SUSTAINABILITY

Dr. John L. Ross, Jr.

INDUSTRIAL PRESS

Industrial Press, Inc.
1 Chestnut Street
South Norwalk, CT 06854
Phone: 203-956-5593
Toll-Free in USA: 888-528-7852
Email: info@industrialpress.com

Author: Dr. John L. Ross, Jr.
Title: The Storeroom Roadmap
Library of Congress Control Number: On file with the Library of Congress.

Printed in the United States of America.

ISBN (print)	978-0-8311-3694-9
ISBN (ePUB)	978-0-8311-9651-6
ISBN (eMOBI)	978-0-8311-9652-3
ISBN (ePDF)	978-0-8311-9650-9

Publisher/Editorial Director: Judy Bass
Copy Editor: Judy Duguid
Compositor: Patricia Wallenburg, TypeWriting
Proofreader: David Johnstone
Indexer: WordCo Indexing Services

books.industrialpress.com
ebooks.industrialpress.com

For Mary Jo

Also by Dr. John L. Ross, Jr.

The Reliability Excellence Workbook: From Ideas to Action

Cover Your A$$ets: Asset Management at Your Place and at Your Pace

Contents

Foreword

About 15 years ago, I was leading manufacturing operations for a successful hydraulic cylinder manufacturing company. We had great people, great processes, and a great product. Our customers loved us, and business was booming. But it was becoming increasingly difficult to execute our plans because of the reliability of our equipment. Our maintenance posture needed some help. So I went looking for assistance and was blessed to meet Dr. John Ross.

I was introduced to John as he taught World Class Maintenance. We were so impressed that we asked John to do more work with us. Over the next few years, he became a valued partner. John was able to guide us through several initiatives. He trained many of our people in world-class maintenance principles and facilitated workshops that elevated our basic equipment care. Our equipment reliability improved, along with employee morale and business results. And if that had been all that John had done, it would have been worth it. But there was more.

John challenged us to rethink storeroom operations. This was something that, quite frankly, wasn't even on our radar. It certainly wasn't something that came to mind when we started our maintenance improvement journey. But little things indicated that we might have some work to do . . . such as seeing eight parts in the inventory system and zero on the shelf. Or thinking, "How can we have all these parts but not the part we need?" Or "Why do I still have this part? That machine hasn't been in the building for 25 years."

So we followed John's guidance. Back then, he was making it up as he went along—the roadmap came later. (Little late, Bro! If I could have bought this book back then, I could have saved $15,000 in consulting fees.) We were able to tap into the wealth of knowledge that John could offer. The storeroom efforts helped us reduce our inventory dollars, increase part availability, improve issuing responsiveness, and positively impact a host of other metrics.

I was able to follow a similar path with John through other manufacturing industries—a diverse group that spanned semi-trailers to paper packaging. The ideas presented in John's books worked every time with very similar results. They are universal principles of maintenance that apply to anyone needing guidance, regardless of industry.

Sometimes John or a member of his team was alongside us to provide guidance. Other times, I would grab my copy of one of John's other books, *The Reliability Excellence Workbook* or *Cover Your A$$ets* to review the principles and move us back to a better road.

Of the three of John's books, I am perhaps most interested in this one. Storerooms are the forgotten "no-man's-land" of maintenance. They are often a chaotic anarchy. You've probably experienced that, or run into the occasional maintenance hoarder who keeps everything "just in case we might need it someday." We discovered tons (literally) of parts in our storerooms kept by "that guy." To be fair, the parts probably didn't weigh tons, although the accumulated dust from several decades put that number over the top. But they took up valuable space, got in the way, and 90% of the time were never needed again. And for the remaining 10%, it was probably a 50/50 chance that the part was fit for use when we did need it. We can all do better.

When John asked me to write the Foreword to his book, I pushed back a little. Forewords are usually written by people who can help the author sell a lot more books; people with names you would recognize and think, "Wow! If *that* guy liked the book, I better buy it." I'm not that guy. I'm just someone who has lived this journey a few times. And every time I've walked this path, I've been grateful for the guidance and the roadmap laid down by Dr. John Ross. These ideas will work for you. I promise (or John will give you your money back).

Brian Lenderink
March 2024

Acknowledgments

When I met the woman who is now my wife, I told her that I thought I had three books in me. After (at the time) 33 years in maintenance and reliability, I had seen some stuff, and I thought I could write about it, and someone might want to read it. This is my third book.

There have been two exceptional women who have made all three books possible, and I would like to acknowledge and thank them both on this page.

The first is my wife, Mary Jo, to whom I have dedicated this work. Mary Jo has been my encourager, my editor, and my comma-cop through all three books. I met Mary Jo in person for the first time on August 5, 2017. That is an important date, because . . .

Two months and two weeks later I met the second woman. My publisher and *the* editor for Industrial Press, Inc., Judy Bass. I met Judy on October 19, 2017, at the SMRP Conference in Kansas City, Missouri. I've shared this story before.

At the time, I was attending the annual conference and presenting at the event as well. I spotted the Industrial Press bookstore across the conference hall and decided to walk over and ask some questions.

Judy was working at the bookstore, as I have known her to do many times over the years since that day. I asked Judy back then, "Who would I need to talk to about writing a book?" Turns out, I was looking at her.

Both Mary Jo and Judy were there for *The Reliability Excellence Workbook: From Ideas to Action* and *Cover Your A$$ets: Asset Management at Your Place and at Your Pace*. And they are still there. Thank you, ladies. As I promised you both, book number three.

The Journey Begins

And will you succeed?
Yes! You will, indeed!
(98 and ¾ percent guaranteed.)
—Dr. Seuss, *Oh the Places You'll Go!*

Author's Note: Read this book carefully. Twice if necessary. Gather your team, contemplate the information, and navigate as necessary. This book will transform not only your storeroom, but your maintenance and reliability effectiveness. 98 and ¾ percent guaranteed.

As a child, I was always fascinated with maps and globes. Old maps, new maps, maps of space, maps of tunnels, maps of the seabed, it didn't matter. To this day, I often sit at my desk eating my lunch and Google Earth some spot on the globe and "fly" around to see what's there. The 1960s and 1970s were the heyday for map lovers. Foldout maps from gas stations were our catnip.

In elementary school, our educational system subscribed to the SRA Reading Laboratory Kit program, or what was called SRA cards (Science Research Associates, Inc.). My favorite subjects were the cards that taught you about maps, directions, navigation, etc. The series of cards on maps was in the advanced section of the SRA program. I had to suffer through spelling, math, grammar, and social studies to get to my quest. It was hard work for a young man, but I was hooked.

What did I learn at such an early age? With a map, you can determine where you are geographically. You can locate points of interest around you, determine obstacles that might be in front of you and impede your journey, or discover some hidden gem that you would otherwise unknowingly travel past. Maps help you to navigate, find the best route, and plan for future efforts and events. A map gives you an unlimited and total accounting of what lies around you and ahead of you. A map opens a world of possibilities if you would only take that first step. If you have a map and know where you are, then you can never be lost.

THE ORIGIN OF THE STOREROOM ROADMAP

I had been a contract consultant for a large maintenance and reliability consulting organization out of Raleigh, North Carolina, for many years. In fact, when my time as a contractor was over, I had become the organization's most *senior* "senior" consultant. I still remain the only contract consultant the organization has ever placed in charge of one of its major business verticals (business streams). I had been asked by the president and later confirmed by the owner of the company to head up the storeroom division. A recently departed employee had been hired to create a suite of storeroom products and services. However, not much progress had been made. When I asked why the company was considering me, a contractor, for this responsibility, the answer from the president and the owner was simply, "Because we know you'll get it done."

The first formal action I took was to revamp the storeroom course. This organization is well known for its public training and has some great products. However, the storeroom course material was old and did not *speak* to the issues and concerns of the day. The materials were dated, and I thought the course just didn't *move* people.

During the course rewrite, headed up by the curriculum development team, we had some revolutionary thinking and conversations. Not all the conversations took place in the conference room; they often spilled out into the hallways and break room. It was during one of these extemporaneous discussions that the Storeroom Roadmap was conceived.

I want to give credit for the idea of the roadmap to an associate of this company, who was more of an IT person but was honestly leaned on heavily by the company president as sort of a "Man Friday." The gentleman's name is Stephen Miller. Stephen was and is extremely bright, and his questions and input were always on point and very much at a level that helps to put things into perspective. Often, at a higher level.

Stephen asked me a very simple question during one of the breaks, and I'm embarrassed to say that I hadn't thought of this question before. Stephen asked, "Is there an order in which you should do things in the storeroom?" I recall asking him what he meant by that, and he said, simply, "Are there processes that make you more effective and processes that make you more efficient?" That was the seed that got the whole thing started. From the tiniest acorn springs the mighty oak.

My part was easy. Divide the processes and practices up between two hemispheres: *effective* and *efficient*. Stephen additionally suggested ranges of *foundation*, *intermediate*, and *advanced*.

The Storeroom Roadmap was generated that very day. Since then, the roadmap has been featured in two best-selling books and hundreds of public and private courses and has been shared around the world. It is a mainstay diagram within the Certified Storeroom Manager/Technician certification program.

HOW THE ROADMAP WORKS

The format of the Storeroom Roadmap has changed, but hasn't changed much since its inception years ago. Stephen had asked if there were some processes and practices that made a storeroom more effective and some that made the storeroom more efficient. That certainly made sense, as did his suggestion that the processes and practices be parsed into foundation, intermediate, and advanced categories. I've added some organizational features along the way in the spirit of continuous improvement.

It should be noted here at the beginning of this book, and should be kept in mind throughout the review of the book's material, that the roadmap is one person's opinion of how the terrain of the storeroom should be laid out and navigated. I humbly suggest that after almost four decades in maintenance and reliability and almost two of those decades consulting in storerooms around the world, I have an educated opinion on how things work best. However, read and study the map with an open mind. Your obstacles may very well be different from what others are facing. Your opportunities, demands, and desires might take you on a different path from the suggested routes on this map. I liken this point to the early days of GPS. When the computer told you to make a turn you knew would take you over a cliff, you didn't make the turn.

This map is a tool. Laid out to help you master and benefit from the 32 major storeroom processes and practices that are applicable to almost every storeroom on earth.

Use it as the tool it is. And don't drive over a cliff.

The Storeroom Roadmap is laid out in a grid of six zones. The 32 processes and practices appear on the map within those six zones. As well, there are other features on the map that are illustrated and will be discussed in context throughout this book.

The six zones are:

1. Foundation-Effective
2. Foundation-Efficient
3. Intermediate-Effective
4. Intermediate-Efficient
5. Advanced-Effective
6. Advanced-Efficient

Figure 1 shows the Storeroom Roadmap in detail (it follows on the next page spread).

Table 1 is a list of the 32 processes and practices identifying their "zone" location along with the corresponding chapter in which they are discussed in detail. I've also included the other features on the map and their reference information. The other features include what I consider essential organizational elements to round out an effective and efficient storeroom operation.

TABLE 1 32 Storeroom Processes and Practices

Element	Activity	Zone	Chapter
Defined Space	Organizational	Foundation-Effective	2
Security	Organizational	Foundation-Effective	2
Physical Organization	Organizational	Foundation-Effective	2
Staffing Roles and Responsibilities	Organizational	Foundation-Effective	2
New Item Setup	Process	Foundation-Effective	2
Order Parts	Process	Foundation-Effective	2
Receiving	Process	Foundation-Effective	2
Issuing	Process	Foundation-Effective	2
Stores Stock Committee	Organizational	Foundation-Efficient	3
Data Scrubbing	Practice	Foundation-Efficient	3
Kitting	Process	Foundation-Efficient	3

Element	**Activity**	**Zone**	**Chapter**
Return to Stock	Process	Intermediate-Effective	4
Return to Supplier	Process	Intermediate-Effective	4
Emergency Procurement	Process	Intermediate-Effective	4
ABC Classification Review	Process	Intermediate-Effective	4
Critical Spares Review	Practice	Intermediate-Effective	4
Cycle Counting	Process	Intermediate-Effective	4
Inventory Accuracy	KPI	Intermediate-Effective	4
Inventory Service Level	KPI	Intermediate-Effective	4
Document Stockouts	Process	Intermediate-Effective	4
Adjust ABC Classification Model	Process	Intermediate-Efficient	5
Critical Spare Parts Algorithm	Practice	Intermediate-Efficient	5
Cycle Counting Criteria	Process	Intermediate-Efficient	5
Optimizing Stock Levels	Practice	Intermediate-Efficient	5
Requisition of Materials	Organizational	Intermediate-Efficient	5
Receiving Purchased Items from Stores	Organizational	Intermediate-Efficient	5
Parts Standardization	Process	Intermediate-Efficient	5
BOMs	Organizational	Intermediate-Efficient	5
Optimized Physical Layout	Organizational	Intermediate-Efficient	5
Special Tools	Practice	Intermediate-Efficient	5
Safety Items	Organizational	Intermediate-Efficient	5
Technical Library	Organizational	Advanced-Efficient	5
Bench Stock	Practice	Intermediate-Efficient	5
Lighting Survey	Organizational	Intermediate-Efficient	5
Last Used Report	Process	Intermediate-Efficient	5
Obsolescence Criteria	Process	Intermediate-Efficient	5
Obsolescence	Process	Intermediate-Efficient	5
Item Substitution	Practice	Intermediate-Efficient	5
Repair or Replace	Process	Intermediate-Efficient	5
Salvage Value	Practice	Intermediate-Efficient	5
Disposal of Scrap	Practice	Intermediate-Efficient	5
Inventory Turn Ratio	KPI	Advanced-Effective	6
Preservation Program	Process	Advanced-Effective	6
RCA—Root Cause Analysis	Organizational	Advanced-Efficient	7
Monthly Budget	Organizational	Advanced-Efficient	7
Stores Issues vs. Budget	Organizational	Advanced-Efficient	7
Delivery	Process	Advanced-Efficient	7
Spare Parts Loan	Practice	Advanced-Efficient	7
VMI/Consignment	Practice	Advanced-Efficient	7

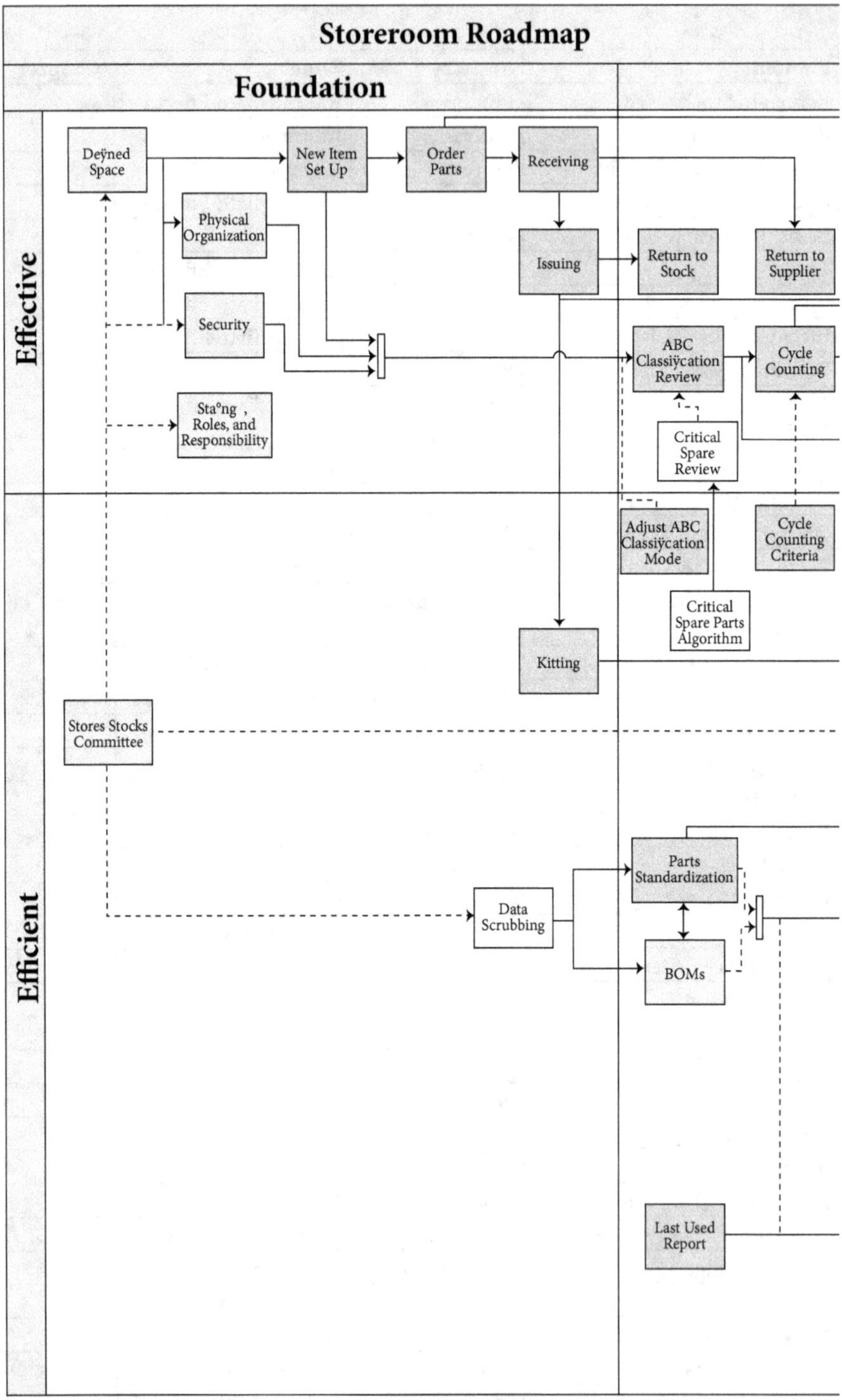

FIGURE 1 The Storeroom Roadmap.

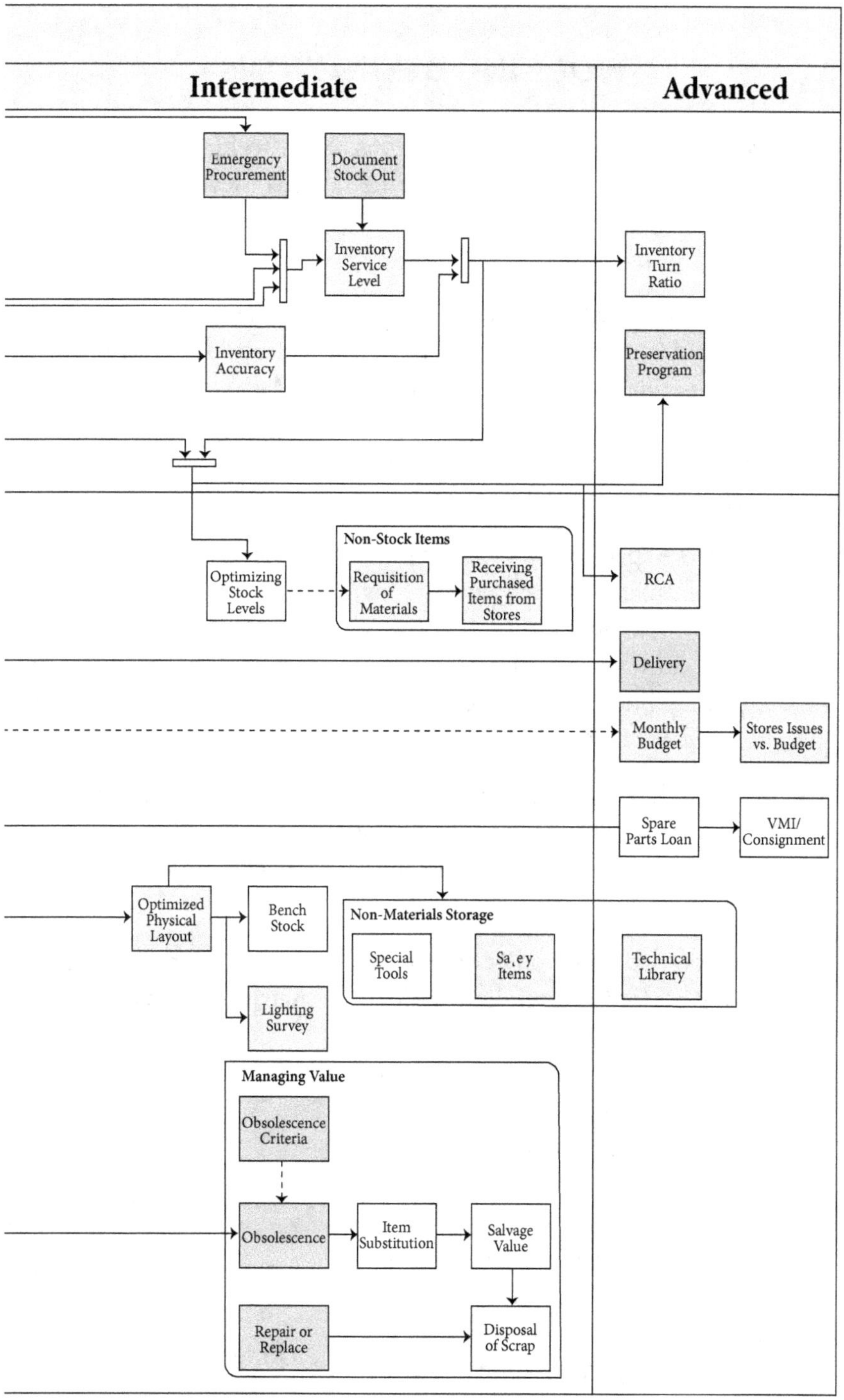

FIGURE 1 The Storeroom Roadmap (*continued*).

WORKING DEFINITIONS

I like to operate with working definitions. Now, to be sure, these aren't likely to be *Webster's* definitions, but rather a "street-level" understanding of what the word or subject might mean. After all, how many of us operate day-to-day quoting *Webster's*?

Processes. More specifically, tactical processes. These are storeroom functions that are usually enacted *no fewer* than 12 times per year. Processes focus on *sustaining* and usually drive leading indicators. Oftentimes, a process will connect two departments (e.g., maintenance-storeroom). These are likely to be the daily activities and the primary functions of the storeroom (*Materials Management for Maintenance*, p. 181).

Practices. More specifically, strategic practices. These are storeroom functions that are usually enacted *no more* than four times per year. Practices focus on *improvement* and usually drive lagging indicators. Oftentimes, a practice will connect several departments (e.g., storeroom-maintenance-finance [think "budget"]). These are typically activities that are enacted quarterly (*Materials Management for Maintenance*, p. 52).

Organizational. These are operating features that are necessary for any functioning entity. The storeroom, like every other segment of business, requires sound management, leadership, and background workings to round out smooth operations.

KPIs/Metrics. A distinction is drawn here because a metric may or may not be a key performance indicator (KPI). A metric, or measure, is simply a value used to show the scale of something. The measured point is just a moment in time, and the value could be used for a myriad of follow-on activities. What makes a metric a KPI is that the measure is tied directly to a stated goal or objective, thus making it "key." A "stated" goal or objective is one that is documented with active effort made to control or improve the value. Figure 2 shows the correlation between a metric and a KPI.

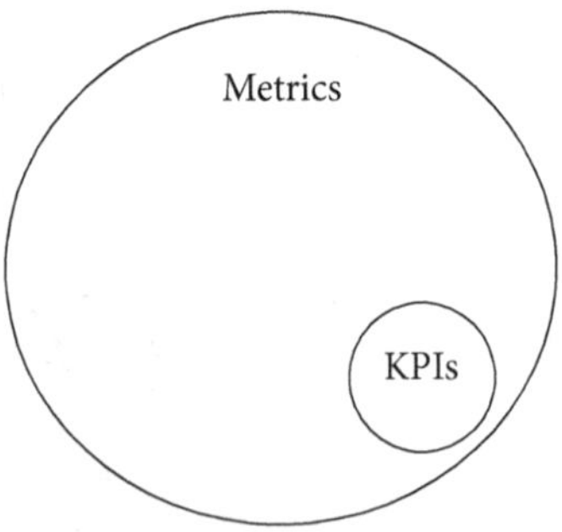

FIGURE 2 Metrics vs. KPIs.

Foundation. Base level. Foundation, or foundational, elements are essential to build upon. You can build on a solid base of strong, fundamental policies, but you'd be well advised not to skimp and shortcut the bedrock of your strategy. The foundation is the bedrock.

Intermediate. Transitional activities. There is a "space" between the foundation and the advanced execution of best practices. A good metaphor for this idea would be a varsity sports program. A college varsity program is the intermediate position between high school and major league. The college level, in sports, is where athletes get high-level, skill-position coaching. In college sports, athletes are no longer only going against cross-town rivals. No. They are competing against the best in the world. This is in preparation for the major league.

Typically, a person doesn't just "arrive" at the major leagues. It is necessary for everyone to transition from foundation through intermediate on the way to the big league.

Advanced. The highest attainable point. For the purpose of the storeroom map, it is greatly advised that organizations start at the foundation and work their way toward advanced. The advanced column is the "heavy lifting" and can often be a source of great frustration. If a company finds itself frustrated in these zones, it is almost 100% due to inadequate attention to all the processes and practices that preceded the point of heartache.

Effective. It worked, likely on the first try and with the desired results. When something is effective, the action you took satisfied the requirement. For example, a preventive maintenance (PM) task is effective if it is executed properly and the component that was the subject of the PM remains operational through the next PM cycle. An activity that is said to be one that makes your operation more "effective" is one that establishes the proper activities to execute at the proper frequency and with the proper amount of scrutiny for continuous improvement purposes. If done right, an effective storeroom process or practice satisfies the requirement.

Efficient. Simply achieving the "mostest" for the "leastest" is my favorite way to explain efficiency. Isn't it the aim of every organization to get the greatest output for the least amount of input? In fact, productivity is the result of dividing output by input. If an activity is efficient, then the least amount of joules of energy was expended in completing it.

MAPPING OUT YOUR JOURNEY

No two storerooms are the same; no two routes on this journey are the same. There is a fair chance that most organizations will have most of these elements already functioning in their storerooms and don't need to *plot* a course as much as they will need to *correct* a course.

Aside from guiding the mature and new storerooms to a level of sustainability and excellence, there are a lot of sensible connections that you might *see* for the first time as you navigate through this roadmap. For example:

- *All* elements flow from the foundational elements encountered in the Foundation-Effective zone.
- Performing the first eight elements detailed on this map at a world-class level won't guarantee a world-class storeroom, but not performing them at a world-class level *will* guarantee you won't have a world-class storeroom. You probably won't even have a *good* storeroom.
- The Order Parts process must have an accompanying Emergency Procurement process.
- The Receiving process must have an accompanying Return to Supplier process.
- The Issuing process must have an accompanying Return to Stock process.
- You can't effectively measure the element of Inventory Turn Ratio without measuring Inventory Service Level and Inventory Accuracy.
- You can't have an Obsolescence process without a way to dispose of components with accompanying Salvage Value and Disposal of Scrap practices.
- And many more connections will be encountered. Just like on a roadmap where you have intersections, roads merging, and cul-de-sacs.

You will find this book to be very helpful in explaining the most basic and necessary elements that virtually all storerooms have in common. This Storeroom Roadmap is laid out in a logical flow, and the manner in which the elements build on the preceding elements will soon make more and more sense. If your storeroom is underperforming, it is a sure bet that there is an element somewhere on the path that is not being executed at the highest level. Instead of addressing it, and addressing it correctly, your team may have chosen to go around it. Does that sound like world class?

CHAPTER 1

THE RULES

Standardized work was never intended by Toyota to be a management tool to be imposed coercively on the workforce. On the contrary, rather than enforcing rigid standards that can make jobs routine and degrading, standardized work is the basis for empowering workers and innovation in the workplace.

—Jeffrey Liker, *The Toyota Way*

The presumption, or is it an *assumption*, is that you are reading this book and will use this book as a reference to make your storeroom better—more effective and more efficient. You are reading this with hopes of contributing to the company's overall strategic initiative to improve asset reliability and, ultimately, the bottom line. Lofty goals with proper motivation.

But there are some rules. If you use this roadmap, and you should, you should comply with some very fundamental yet critical rules. If not, you will be more frustrated with your storeroom operation than you are today. Guaranteed.

Follow these rules, and you'll be well on your way to an exceptional storeroom operation.

RULE #1—PROCESS GUIDES

Without exception, all 32 of the storeroom processes and practices must have an established governance policy. I suggest that this policy should be in the form of a process guide. A process guide provides the general guidance necessary for the sustainable execution of the process. The sustainable execution of a process nets a desired result every time!

Alain Pellegrino, in his *Reliability Workshop for Executives,* identified having "Standard Operating Procedures, in place and followed" as the fourth of seven signs of a reliable plant (p. 39). How can an exceptionally effective and efficient process be repeated at the highest level if it isn't detailed out in an operating procedure? It can't.

This probably sounds like a lot of work, and it is. However, I suggest it doesn't become "less" work the longer you wait to do it. Can you honestly think of any process, the result of which is so essential to success, where the directions *shouldn't* be documented, trained to, and judiciously followed? Let alone improved upon through the methodical process of continuous improvement? I can't think of a single important operation in the storeroom that shouldn't follow Rule #1.

There is a great analogy to the importance and the doctrine of having a process and following the process. This analogy, once again, centers on sports: How many touchdowns can a team expect to score if each member on the offensive team is running his own play? The answer is quite obvious—zero. You will not likely score a point if everyone on the offensive side of the football is running the play he wants to run. If, however, we all subscribe to the play that is called (you know, the one that we've practiced), it is more likely that we'll be successful. *Counter note:* We hope the defense is less disciplined and those 11 guys are doing whatever they want to do.

Process guides have a suggested structure that will lend a greater sense of clarity and formality. The key elements of a standard process guide are:

Business Process Description. In this section of the process guide, the reason for establishing the directive is explained, and an accounting of the major "players" and their role in the process is laid out in general terms.

Applicability. The section on applicability is necessary to document whom the process guide is meant to help in terms of the owner of the process and a list of those affected by or who can influence the process.

Business Process Overview. This section highlights the necessary inputs, desired outputs, and associated "touchpoints," as well as outlining the key design decisions, requirements, and expectations. A "touchpoint" is the intersection of the process that is being defined with another process. For example, when creating the flow for the Ordering process, it would be necessary to "touch," or intersect with, the Purchasing process.

Inputs are the anticipated actions that will enter into the process being defined. Outputs are those, understandably, that are the anticipated outcomes, or results of executing the process.

Key design decisions, requirements, and expectations can be easily explained as an overview of the process itself. For example, for the Ordering process, it is understood that the requester must provide correct and complete information and that items are only purchased from approved vendors. For a short explanation of this section, consider documenting what must "go right" for the process to work smoothly.

Business Process Workflow. This is simply a flowchart of the process. I suggest using what are referred to as "swim lanes," with those agencies that are active in the process listed on the left-hand side of the flowchart, and those activities that the agencies are responsible for shown in their "lane" as you move to the right within the flowchart. The advantage of this design is that it provides a clear understanding of *who* is responsible for *which* activity.

In your flowcharts, use common flowchart symbols as shown in Figure 1-1. Figure 1-2 is an example of a flowchart with defined swim lanes.

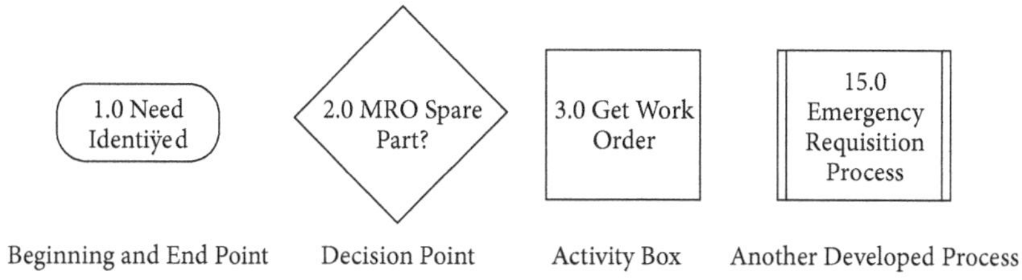

FIGURE 1-1 Common flowchart symbols.

There is a little-known secret to flowcharts that, if executed consistently, can make your work more viable and easier to follow. For each of the symbols that are on your flowchart, number them in the order in which you want the process to flow. In this manner, you can refer to an activity by number, and this number will correspond with other sections of the process guide. But more importantly, you have the ability to refer to the simplicity of a process if it is executed correctly. For example, you could say, "If we do this process correctly, it only takes 14 steps. Otherwise, it could take us all week!"

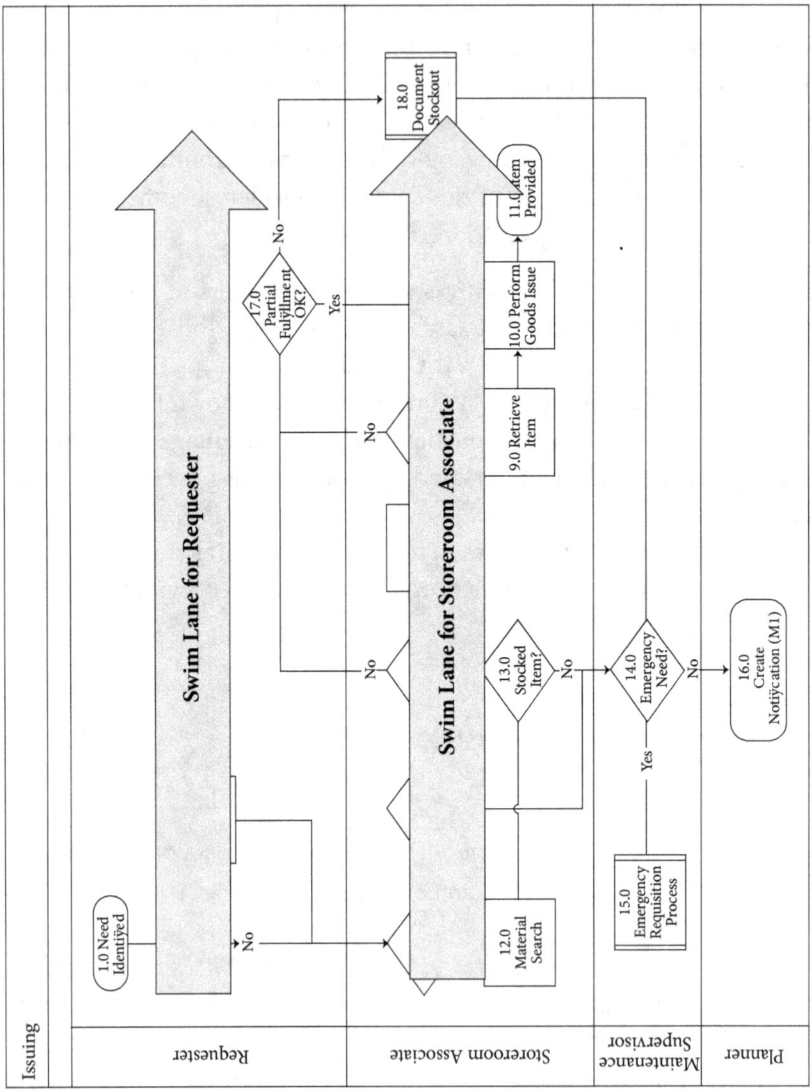

FIGURE 1-2 Flowchart with swim lanes.

This is a great time to invoke a wonderful passage from Tom Moriarty's excellent book, *The Productive Leadership System*. In his book, Tom defines "proactive" as this: "Proactive—I determine the outcomes by influencing the future" (p. 125). Is there a better way to influence the future than to draw out the ideal process flow in steps numbered in the order they are to be executed? A well-defined process, one that was developed by engaging the process users, gives the organization hope for success. Better, it gives the users a plan!

"Hope springs eternal," said Alexander Pope. It is the nature of humans. Is there any more empowering thought, to go along with Moriarty's definition of proactive, than Tim Stanley's explanation in the *Tulsa World* newspaper that "Hope Science defines hope as the understanding that not only can the future be better but that I as an individual have the power to make it so through the active role I play" (p. A1)? As leaders, let's commit ourselves to proactively leading the way, and instilling the hope that we can make a difference by taking an active role. Engage the workforce and create a process that works every time.

RACI. The RACI (Responsible, Accountable, Consulted, and Informed) chart indicates the type and level of involvement required from designated roles in the organization for each detailed process activity. The RACI chart activities are laid out as the workflow flowchart in a precise numerical sequence. There are four levels of engagement in the RACI chart:

1. **R**—Responsible. The letter R is used to signify who is responsible for executing the activity on the RACI chart. There should be only one R per activity.
2. **A**—Accountable. The letter A is used to signify on the RACI chart who is ultimately *accountable* for accomplishing the activity. This is typically the supervisor of the person with the R by their name for that activity. Although it seems like you may have more than one boss, you only have one. As a result, there should only be one A per activity.
3. **C**—Consulted. The letter C is meant to be used in two different ways. The first use denotes the actors who were instrumental in detailing specifically how that activity will be executed. The second use denotes those actors you should consult with when that particular activity is not being executed well or consistently. The people help establish how that particular task is to be done and

can then help determine why it isn't being done correctly. There can be more than one C per activity.

4. **I**—Informed. The letter I is necessary to identify others who need to be made aware of the process activities. Since it is impractical to think that everyone can be consulted on each task needed for all 32 processes and practices, a system should be in place to educate and inform pertinent individuals on how the process steps work. The I simply answers the question, "Who else needs to know about this?" There can be more than one I per activity.

Figure 1-3 is an example showing what a completed RACI chart might look like.

Index	**Activity**	**Reliability Engineer**	**Planner**	**Storeroom Supervisor**	**Storeroom Associate**	**Requester's Supervisor**	**Requester**	**Maintenance Manager**	**Maintenance Supervisor**
1	Need Identified					A	R		
2	MRO Spare Part?					A	R		
3	Get Work Order					A	R		
4	Part Number Known			A	R				
5	Cataloged Item?			A	R				
6	Requested Qty Available?			A	R				
7	Locate Item			A	R				
8	Requested Qty on Shelf?			A	R				
9	Retrieve Item			A	R				
10	Perform Goods Issue			A	R				
11	Item Provided			A	R				
12	Material Search			A	R				
13	Stocked Item?			A	R				
14	Emergency Need?							A	R
15	Emergency Requisition Process								
16	Create Notification (M1)	A	R						
17	Partial Fulfillment OK?					A	R		
18	Document Stockout								

FIGURE 1-3 RACI chart.

Activity Detail. This last section of the process guide is used to detail the steps and activities necessary to complete the process. It expands on the information contained in the workflow flowchart. The activities should be detailed by the numbering sequence established in the flowchart and the RACI chart.

A Process Guide Example. The following is a well-documented process guide for you to use as an example. This guide is for the storeroom process of Issuing. Read through this example and consider the value of having such documented governance in your organization. The existence and adherence to formally developed and documented processes is key to your storeroom's success and sustainability. This process guide was first printed in *The Reliability Excellence Workbook: From Ideas to Action* (Ross, pp. 139–146).

ISSUING

Business Process Description

The purpose of the Issuing process is to set forth guidelines for issuing items from the Storeroom. All items issued from the Storeroom inventory must be issued to a valid work order and tracked against the work order's financial authority for accountability. Item issues will only be made directly to a valid work order with or without a reservation for the item and will not be made directly to a cost center.

The primary roles involved in this process are:

- The Requester—makes request for item(s), takes possession of item(s)
- The Storeroom Associate—confirms the presence of a valid work order, retrieves item(s), documents stockout occurrences, issues item(s)
- The Maintenance Supervisor—determines if the need is an emergency, executes the Emergency Requisition process
- The Planner—creates a notification

Applicability

The Issuing process is owned by the Maintenance Department with consultation from Production as the primary end users. The target audience for distribution of this process includes, but is not limited to, the following departments: Storeroom, Production, and Procurement.

This process is not intended to be used for issuing items for capital projects.

This process is to be used for the issuance of all items issued, tracked, and controlled by the Storeroom, including parts, materials, and consumables related to plant Maintenance and Production. This process is not intended to be used for tools or everyday consumables utilized within the administrative or office setting (e.g., office supplies).

Business Process Overview

This section highlights the necessary inputs, desired outputs, and associated touchpoints (where the Issuing process relies on or interfaces with other processes), and outlines the key design decisions, requirements, and expectations.

The following **input** is necessary for completing the Issuing process:

- Valid work order

The following **outputs** are the expected result of the Issuing process:

- Item issued to the Requester
- Adjustment of on-hand inventory to reflect the issuance of item(s)
- Notification that the item is not available
- Work notification, M1

The Issuing process has the following **touchpoints** with other processes:

- Execution of the Issuing process could involve the execution of the Document Stockout process
- Execution of the Issuing process could involve the execution of the Emergency Requisition process

The following are the **key design decisions, requirements, and expectations** for the completion of this process:

- Proper execution of the Issuing process aids in keeping inventory accurate and helps when researching root cause issues (for history)
- No items are issued without a valid work order
- Items that are to be issued under this process must first be items that are, in fact, stocked items to begin with

Business Process Workflow

The following flowchart gives a high-level overview of the process flow for Issuing. It illustrates necessary inputs, desired outputs, and any touchpoints (where the Issuing process relies on or interfaces with other processes).

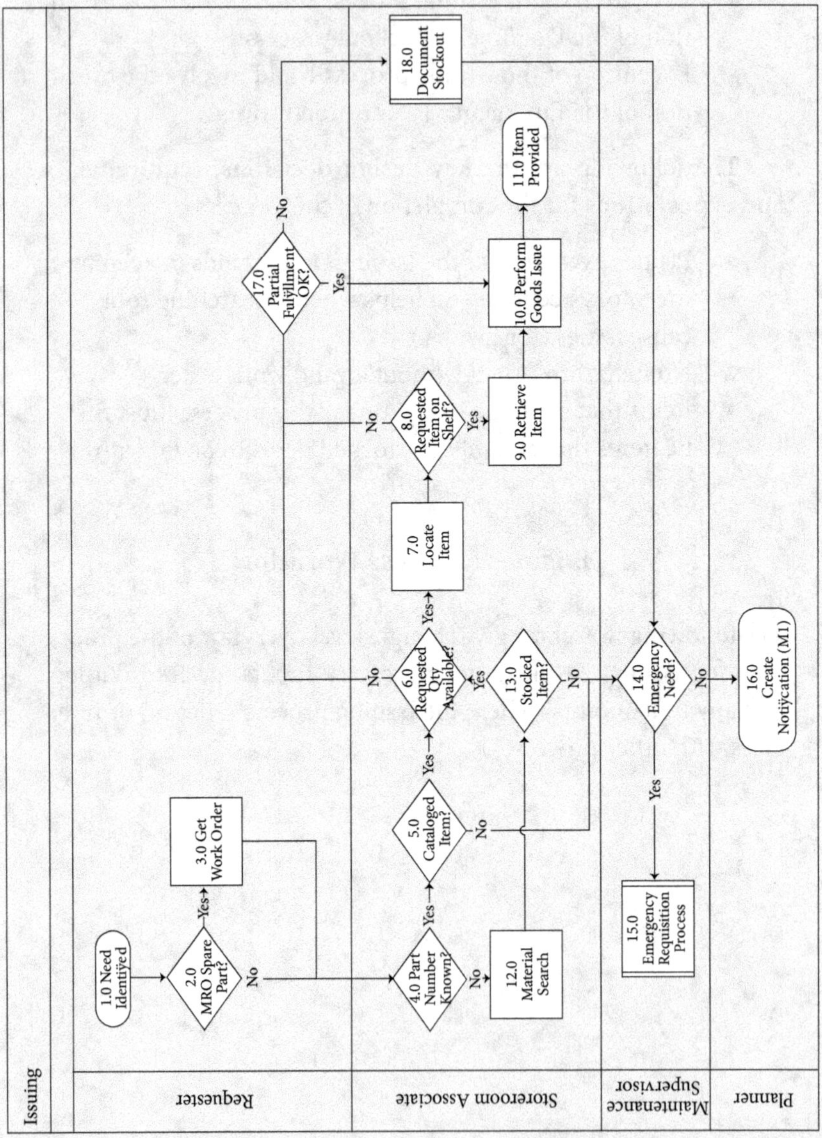

Issuing
Requester
Storeroom Associate
Maintenance Supervisor
Planner
1.0 Need Identiÿed
2.0 MRO Spare Part?
Yes
No
3.0 Get Work Order
4.0 Part Number Known?
Yes
No
5.0 Cataloged Item?
Yes
No
6.0 Requested Qty Available?
Yes
No
7.0 Locate Item
8.0 Requested Item on Shelf?
Yes
No
9.0 Retrieve Item
10.0 Perform Goods Issue
11.0 Item Provided
12.0 Material Search
13.0 Stocked Item?
Yes
No
14.0 Emergency Need?
Yes
No
15.0 Emergency Requisition Process
16.0 Create Notiÿcation (M1)
17.0 Partial Fulÿllment OK?
Yes
No
18.0 Document Stockout

RACI

The RACI (Responsible, Accountable, Consulted, and Informed) chart indicates the type and level of involvement required from designated roles in the organization for each activity of the Issuing process.

Index	Activity	Reliability Engineer	Planner	Storeroom Supervisor	Storeroom Associate	Requester's Supervisor	Requester	Maintenance Manager	Maintenance Supervisor
1	Need Identified					A	R		
2	MRO Spare Part?					A	R		
3	Get Work Order					A	R		
4	Part Number Known			A	R				
5	Cataloged Item?			A	R				
6	Requested Qty Available?			A	R				
7	Locate Item			A	R				
8	Requested Qty on Shelf?			A	R				
9	Retrieve Item			A	R				
10	Perform Goods Issue			A	R				
11	Item Provided			A	R				
12	Material Search			A	R				
13	Stocked Item?			A	R				
14	Emergency Need?							A	R
15	Emergency Requisition Process								
16	Create Notification (M1)	A	R						
17	Partial Fulfillment OK?					A	R		
18	Document Stockout								

Activity Detail

The following section details the activities necessary to complete the Issuing process. This expands the information contained in the workflow.

1. **Need Identified**

   The Issuing process begins when the Requester identifies a need for a particular item.

   Activity complete—Go to activity 2, MRO Spare Part?

2. **MRO Spare Part?**
The Requester, possibly in consultation with others, will determine if the needed item is an MRO spare part. A spare part requires a work order so the withdrawal from stores can be charged to an asset.
   Yes—Go to activity 3, Get Work Order
   No—Go to activity 4, Part Number Known?
3. **Get Work Order**
The Requester must obtain a work order to be presented to the Storeroom. Preferably, the valid work order will list a part number to assist the Storeroom Associate in locating the required item. In the event the Requester does not have a valid work order, it is the Requester's responsibility to obtain one before the Storeroom Associate can issue any item(s).
   Activity complete—Go to activity 4, Part Number Known?
4. **Part Number Known?**
The Storeroom Associate will determine if the item being requested has a known part number. A part number could indicate that the item is stocked or, at the very least, can be ordered to specification. Otherwise, additional steps are necessary to locate the item.
   Yes—Go to activity 5, Cataloged Item?
   No—Go to activity 12, Material Search
5. **Cataloged Item?**
The Storeroom Associate will determine if the item being requested is a cataloged item. A cataloged item is one that has a known part number and all associated information to order but may not really be an on-hand stocked item.
   Yes—Go to activity 6, Requested Qty Available?
   No—Go to activity 14, Emergency Need?
6. **Requested Qty Available?**
When attempting to issue an item to a work order, the Storeroom Associate will validate via the inventory management system if there is sufficient quantity on hand to satisfy the request.

If sufficient quantity is shown on hand, the MRO Storeroom Associate will obtain and issue the item.

In the event there is insufficient quantity showing on hand, the MRO Storeroom Associate will inform the Requester of the potential problem with immediately filling the request.

In the event there is an on-hand quantity, but it is insufficient to satisfy the request, the MRO Storeroom Associate will offer to partially fill the request but leave the decision to do so up to the Requester. If the Requester chooses to take the partial issuance, the Storeroom Associate will issue the on-hand inventory.

Yes—Go to activity 7, Locate Item

No—Go to activity 17, Partial Request Fulfillment OK?

7. **Locate Item**

The Storeroom Associate will locate the item being requested.

Activity complete—Go to activity 8, Requested Qty on Shelf?

8. **Requested Qty on Shelf?**

In the event the item or a subset of the quantity is not the amount needed, that it might be different than what was initially listed in the inventory system previously, the Storeroom Associate will inform the Requester.

Yes—Go to activity 9, Retrieve Item

No—Go to activity 17, Partial Fulfillment OK?

9. **Retrieve Item**

The Storeroom Associate will retrieve the item to be issued from the location shown in the inventory management system.

Activity complete—Go to activity 10, Perform Goods Issue

10. **Perform Goods Issue**

The Storeroom Associate will record the item's removal from stock in the inventory system.

Activity complete—Go to activity 11, Item Provided

11. **Item Provided**
The Storeroom Associate will provide the item to the Requester.
Activity complete—This completes the Issuing process.
12. **Material Search**
In the event the part number is not known, the Storeroom Associate and the Requester will work to locate the item in the inventory system or by other available means.
Activity complete—Go to activity 13, Stocked Item?
13. **Stocked Item?**
If located, or not, it is the responsibility of the Storeroom Associate to determine if the item being requested is even a cataloged item, and within that description, if it is a stocked item (indicating that an inventory level is expected).
Yes—Go to activity 6, Requested Qty Available?
No—Go to activity 14, Emergency Need?
14. **Emergency Need?**
The Maintenance Supervisor, in consultation with the Requester, will determine if the need for the item is an emergency.
Yes—Go to activity 15, Emergency Requisition process
No—Go to activity 16, Create Notification
15. **Emergency Requisition Process**
The Emergency Requisition Process is a separate process; please see the relevant documentation for the actions required.
Activity complete—This ends the Issuing process
16. **Create Notification**
The Planner will create a notification for M1 jobs, listing the needed items.
Activity complete—This ends the Issuing process.
17. **Partial Fulfillment OK?**
If there is an on-hand quantity but it is insufficient to satisfy the request, the Storeroom Associate will offer to partially fill the request but leave the decision to do so up to the Requester. If the Requester chooses to take the partial

issuance, the Storeroom Associate will issue the on-hand inventory.

Yes—Go to activity 10, Perform Goods Issue

No—Go to activity 18, Document Stockout

18. Document Stockout

Document Stockout is a separate process; please see the relevant documentation for actions required.

Activity complete—Go to activity 14, Emergency Need?

This completes the Issuing process.

Before we leave Rule #1, let me suggest a method for creating these process guides: Do not create process guides unilaterally. That is to say, don't sit in your office and sketch out a flowchart by yourself. This is a great opportunity to engage your workforce by forming a small team and tapping into their creativity to formulate a process flow and procedural steps that they would be willing to execute. You facilitate this process guide development to shepherd the team toward understanding and adopting best practices. You don't want your team to create a process that doesn't align with generally accepted good practices. You definitely don't want them to have process requirements they do not intend to follow.

RULE #2—BILL OF MATERIALS

Just like Rule #1, Process Guides, this rule covering the bill of materials (BOM) is meant to be followed. I can absolutely guarantee that failure to subscribe to these rules with monk-like discipline will result in inconsistency, failure to sustain, and rapid hair loss. Believe me.

Let's first understand what a BOM is. A BOM is a list of the parts that make up an asset. As I documented in *Cover Your A$$ets: Asset Management at Your Place and at Your Pace,* "We likely come across a BOM as an exploded-view diagram that is accompanied by several pages of callouts for each specific item in the diagram" (p. 135). Recall those 1970s-era Chilton's auto repair manuals. There were typically exploded-view/callout images such as those shown in Figures 1-4 and 1-5.

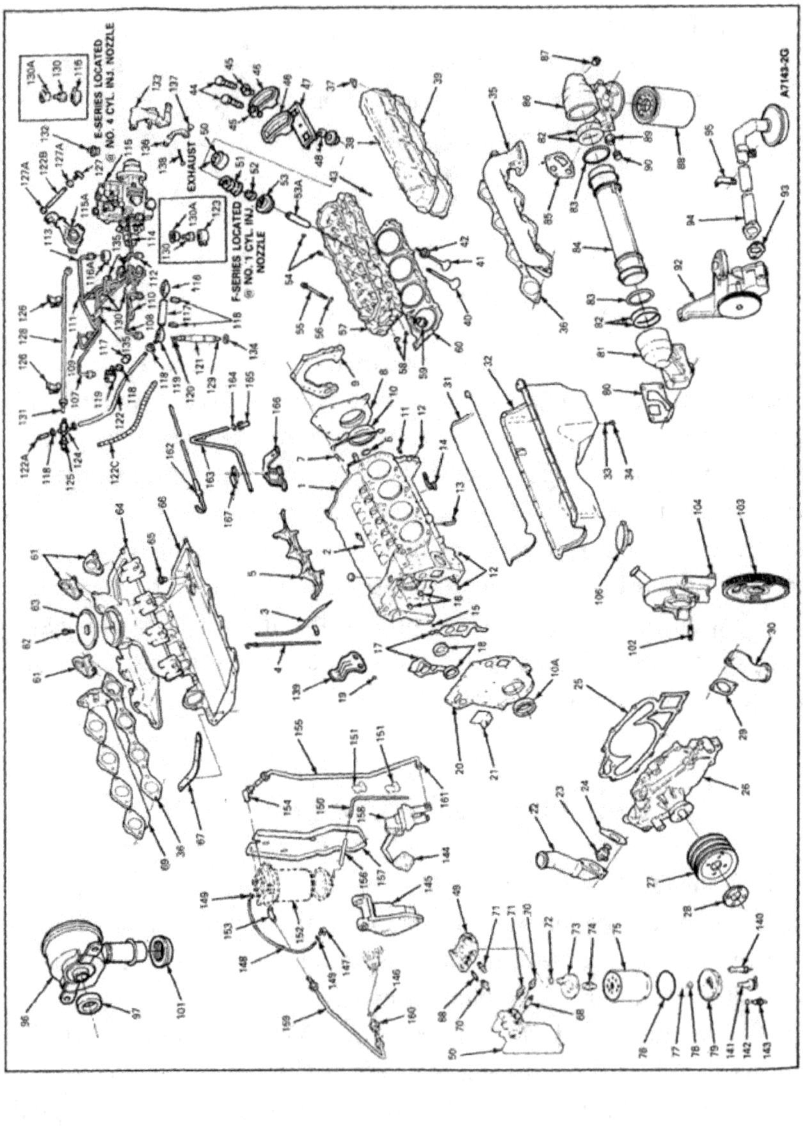

FIGURE 1-4 Exploded-view diagram.

REF. NO.	BASIC PART NO.	DESCRIPTION
1	6009H	Cylinder Block Assy.
2	6C329A	Guide, Tappet
3	6754	Tube Assembly, Oil Level Gauge (F-Series)
4	6750	Oil Level Gauge (F-Series)
5	6C330A	Retainer, Tappet Guide
6	6026A	Plug, Engine (1-1/2" O.D.)
7	6D083A	Gasket, Rear Cover
8	6L080A	Cover Assembly, Engine, Rear
9	6A369A	Adapter, Flywheel to Transmission
10	6701A	Rear Oil Seal, Crankshaft
10A	—	Front Oil Seal
11	6B041B	Dowel Pin, Fly Wheel Adapter
12	87614S	Pipe Plug, 1/8 NPTF
13	6C327A	Piston Cooling Jet
14	6A051A	Heater Assembly, Block
15	6B041A	Dowel Pin, Front Cover Plate
16	6026E	Cup Plug
17	6020A	Gasket, Front Cover Plate
18	6A251A	Bearing Kit, Camshaft
19	6A628A	Ball, Oil Indicator Hole 11/32"
20	6B070A	Plate, Front Cover
21		Indicator, Timing (Part of Front Cover)
22	8592G	Connection, Water Outlet
23	8575	Thermostat
24	8255A	Gasket, Water Outlet
25	8507A	Gasket, Water Pump
26	8501D	Water Pump
27	8509D	Pulley, Water Pump
28	8546A	Spacer, Fan
29	8255A	Gasket, Water Inlet
30	8592D	Connection, Water Inlet
31	D6AZ-19562-A	RTV Sealant
32	6675C	Oil Pan
33	6734A	Gasket, Oil Pan Drain
34	6730A	Plug, Oil Pan Drain
35	9431B	Manifold, Exhaust, Left
36	9448A	Gasket Exhaust Manifold
37	6A532A	Washer, Valve Cover
38	6582C	Valve Cover
39	6584A	Gasket, Valve Cover
40	6507D	Valve, Intake (8)
41	6505	Valve, Exhaust
42	6057B	Insert, Exhaust Valve Seat
43	6026F	Plug, Ball Type 13/32" (8)
44	—	Bolt, Valve Lever and Washer

REF. NO.	BASIC PART NO.	DESCRIPTION
45	—	Post, Valve Lever
46	—	Lever, Valve
47	—	Retainer, Valve Lever Post
48	—	Lock, Valve Spring Retainer
49	6514A	Retainer, Valve Spring (36)
50	—	Shield, Oil (Exhaust)
51	6513B	Spring, Valve, with Damper (16)
52	6571A	Seal, Valve Stem-Intake (8)
53	6K533A	Rotator, Assembly Valve (16)
53A	—	Guide, Valve (Service)
54		Plug, 1/2 inch NPTF (4)
55	6065A	Bolt Cylinder Head (34)
56	6L015A	Washer, Cylinder Head Bolt (34)
57	6049A	Cylinder Head Assembly (2)
58	6026B	Plug, 1/4 Inch
59	6057A	Insert, Combustion Chamber (8)
60	6051B	Gasket, Cylinder Head (2)
61		Eye, Lifting (3)
62	9C629A	Insert, Bolt Thread-Air Cleaner Stud
63	9F460A	Screen, Intake Manifold
64	9424B	Manifold, Intake
65	9A450A	Drain Plug, Valley Pan
66	9439B	Gasket and Valley Pan
67	9B470A	Strap, Valley Pan
68		Fuel Priming Valve and Cap
69	9430A	Manifold, Exhaust, Right
70		Continuous Vent with Check Valve
71		Vacuum Switch (Fuel Filter Element Replacement Indicator)
72		Fuel Heater O-Ring
73		Fuel Heater
74		Threaded Insert
75		Fuel Filter Element
76		Drain Bowl O-Ring
77		Drain Valve Stem Cap
78		Drain Valve Seal
79		Water Separator Drain Bowl
80	6A636A	Gasket, Oil Cooler, Front Header
81		Header, Oil Cooler, Front
82	6K649A	O-Ring, Oil Cooler (2)
83	6C610A	O-Ring, Oil Cooler (2)
84	6A642A	Cooler, Oil
85	6A636B	Gasket, Oil Cooler, Rear Header
86	6881B	Header, Oil Cooler, Rear
87	6K862A	Plug, 1/4-Inch
88	6731A	Oil Filter

REF. NO.	BASIC PART NO.	DESCRIPTION
89		Plug, 1/2 Inch
90		Plug, 1/2 Inch
92	6600	Oil Pump Assy.
93	6626	Gasket, Oil Pick-Up
94	6622	Pick-Up Tube
95	6A661	Bracket, Oil Pick-Up
96	6A665	CDR Valve
97	6A892	Seal Ring, CDR Valve
100	6758	Crankcase Vent Tube
101	6769	Grommet, Valley Cover
102	9F733	Mounting Stud, Injection Pump
103	9A546	Drive Gear, Injection Pump
104	9C516	Adapter Housing, Injection Pump
106	6766	Cap, Oil Filler
107	9A555H	Pipe w/Nuts Pump to Cyl. 8
108	9A555G	Pipe w/Nuts Pump to Cyl. 7
109	9A555F	Pipe w/Nuts Pump to Cyl. 6
110	9A555E	Pipe w/Nuts Pump to Cyl. 5
111	9A555D	Pipe w/Nuts Pump to Cyl. 4
112	9A555C	Pipe w/Nuts Pump to Cyl. 3
113	9A555B	Pipe w/Nuts Pump to Cyl. 2
114	9A555A	Pipe w/Nuts Pump to Cyl. 1
115	9A543	Injection Pump
115A	—	Valve, Vacuum Modulator (Auto. Trans.)
116	—	Fuel Return Tee
116A	—	Elbow, Fuel Return (F-Series)
117	—	Hose
118	9B255	Clip
119	9A584	Fuel Return Tee
120	87032-S92	O-Rings
121	9E527	Injection Nozzle Holder
122	—	Fuel Return Hose
122A	—	Hose
122B	—	Hose, Pump to Fuel Return Tube
122C	—	Guard, Rear Fuel Return Hose
123	—	Fuel Return Tee (E-Series)
124	9F734	Fuel Return Junction Fitting
125	—	Nipple, Fuel Return
126	9N659	Clamp
127	9F736	Elbow
127A	—	Clip
128	9D308	Tube
129	—	Nozzle Tip
130	—	Sensor, Fuel Line Pressure
130A	—	Cover
131	9C387	Sleeve Seal, Fuel Return (2)

REF. NO.	BASIC PART NO.	DESCRIPTION
132	9E939	Temperature Switch
133	12B526	Bracket and Solenoid-Fast Idle
134	—	Gasket Nozzle (8)
135	9N653	Clamp
136	9F541	Kickdown Lever (Auto. Trans.)
137	9D927	Screw, Kickdown Lever
138	9F539	Adjusting Screw (Kickdown Lever)
139	6786	Bracket, Oil Level Tube Support (F-Series)
140		Vent/Valve Assembly
141		Manual Drain Valve
142		Water Sensor O-Ring
143		Water Sensor Probe
144		Fuel Pump Supply Gasket
145		Alternator Bracket
146		Sealing O-Ring
147		Fuel Return Tee (At Nozzle)
148		Hose, 3/16" ID x 10" Long
149		Hose Clip
150		Water Drain Tube
151		Drain Tube Clamp (2)
152		Fuel Filter/Water Separator Element
153		Elbow
154		Elbow, Fuel Supply Pump to Filter Header
155		Fuel Pump to Fuel Header Tube (With Two Nuts and Two Sleeves)
156		Hose, 3/16" x 2-5/16" Long
157		Fuel Filter Header Mounting Bracket
158		Fuel Supply Pump
159		Filter to Injection Pump Tube (With Two Nuts and Two Sleeves)
160		Connector Fitting
161		Inverted Flare Tube Nut
162		Oil Level Gauge — E-Series
163		Tube Assembly, Oil Level Gauge — E-Series
164		O-Ring, Oil Level Gauge — E-Series
165		Oil Level Gauge Tube, Lower — E-Series
166		Bracket, Oil Level Gauge Tube — E-Series
167		Retainer, Oil Level Gauge Tube — E-Series

FIGURE 1-5 Callouts.

The true breadth and value of an accurate and complete bill of materials is starkly apparent when you consider that almost all organizational agencies are affected (positively or negatively) by the bill of materials, as reflected in Figure 1-6.

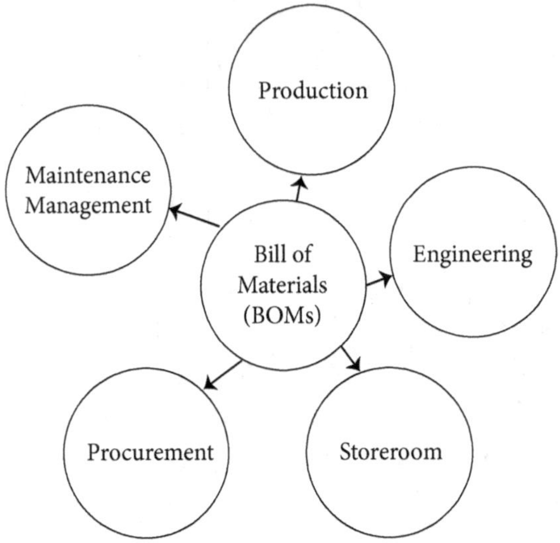

FIGURE 1-6 BOM impact.

Now that we can fully appreciate the multivalue of a thorough bill of materials, the two parts of Rule #2 become clearer and more urgent.

The first part simply states that only items appearing on an asset's BOM will be considered for stocking or registering a stock number. Not every component on a BOM will be stocked, but only the items that appear on the BOM will be considered. Simple.

The second part logically follows the first. Every item in the storeroom categorized as a spare part will be a legitimate part listed on an existing asset's BOM. The storeroom should not and will not stock spare part inventory items that literally don't belong to any active equipment in the facility. This is applicable for every type of maintenance storeroom—manufacturing, pharmaceutical, petrochemical, aerospace, facilities, etc. You document this in the Item Master Data table for the storeroom inventory. There is a field that might be labeled as "where used" and must have a legitimate plant (or facility) asset listed. Sometimes you must verify that the part truly belongs to that asset. You shouldn't be required to challenge those requesting that items be placed into stock, but sometimes you do. If you've spent any time working in a storeroom, you know that people can suggest just about anything for an inventoried item.

If the part does not currently "go" to any equipment in the facility, it should not be stocked in the storeroom. The associates in the storeroom don't need practice in handling items that are not relevant to the operation of the facility.

There is a minor metric that some subscribe to for measuring one of the storeroom's performances. This metric is *storeroom transactions*. The Society for Maintenance & Reliability Professionals (SMRP) defines storeroom transactions in its *SMRP Best Practices* as:

> Any materials management activity that results in the physical handling of an inventory item (stock or non-stock) or direct purchased item that results in the exchange of data with the storeroom inventory management system. Inventory transactions occur any time an item is "touched" either physically or electronically (e.g., a pick list with ten items picked would equal ten transactions). Inventory transactions include receiving, stocking, adding, picking, kitting, staging, issuing, delivering, returning, adjusting, counting inventory stock items, EOQ analysis, etc. (p. 274)

SMRP Best Practices indicates that the best-in-class number of transactions that a storeroom associate can fulfill per day is 100–140 (p. 275). How many of those transactions should be taken up with components that don't fit anywhere or belong to anything? I suggest zero.

Best practice estimates indicate that an organization should have a thorough BOM on 98% of its assets, if not higher. Creating a bill of materials from a blank slate is a daunting task. If you find yourself with an anemic BOM registry, you can consider yourself to be in good company. This is singularly the greatest shortcoming that keeps most storerooms and maintenance operations from reaching the highest level of performance. Most people simply can't be bothered.

It is necessary, and it is a rule. If this is a needed step in your reliability effort, I strongly suggest that you form a small team and charter the team members to create a BOM on your active assets. When doing so, your team should follow the principles laid out in ISO Standard 14224, "Petroleum, petrochemical and natural gas industries: Collection and exchange of reliability and maintenance data for equipment."

This standard ideally lays out the way equipment can be parsed, examined, and "put back" together in a hierarchy that supports the systematic development of a BOM. I find it to be the most basic and straightforward approach to building a bill of materials. Figure 1-7 is taken from ISO 14224 (p. 17) and is an example of how a complex system can be parsed into a hierarchy of components. This is the boundary diagram for a pump.

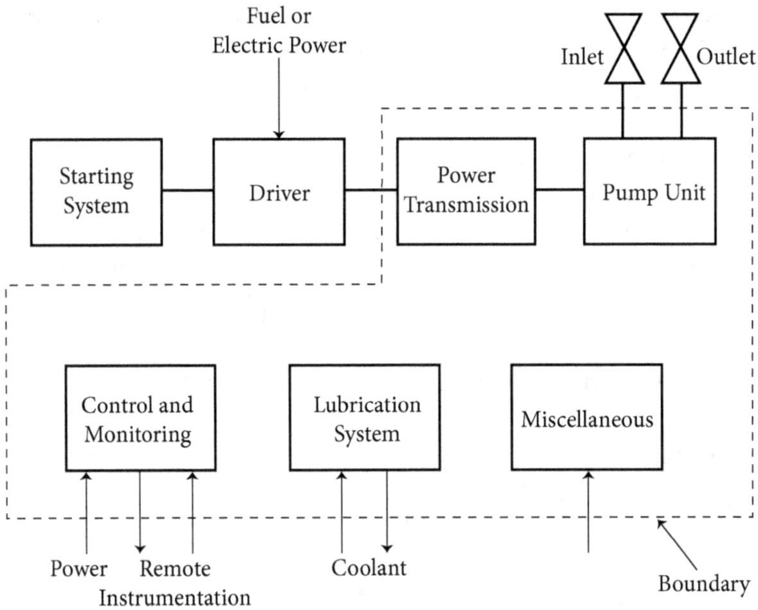

FIGURE 1-7 Example of a boundary diagram.

Figure 1-8 is a continuation of excerpts from ISO 14224 (p. 18) and is of the hierarchy of taxonomy.

Years ago, I created a BOM workbook for a client using ISO 14224. We used it many times, slowly and methodically improving it with each use. I use this BOM workbook every time I'm asked to help start a BOM effort. Use ISO 14224 to construct a truly valuable bill of materials for all your assets. You will be happy that you have this standard to relate to, and you can better compel others to subscribe to a globally recognized methodology.

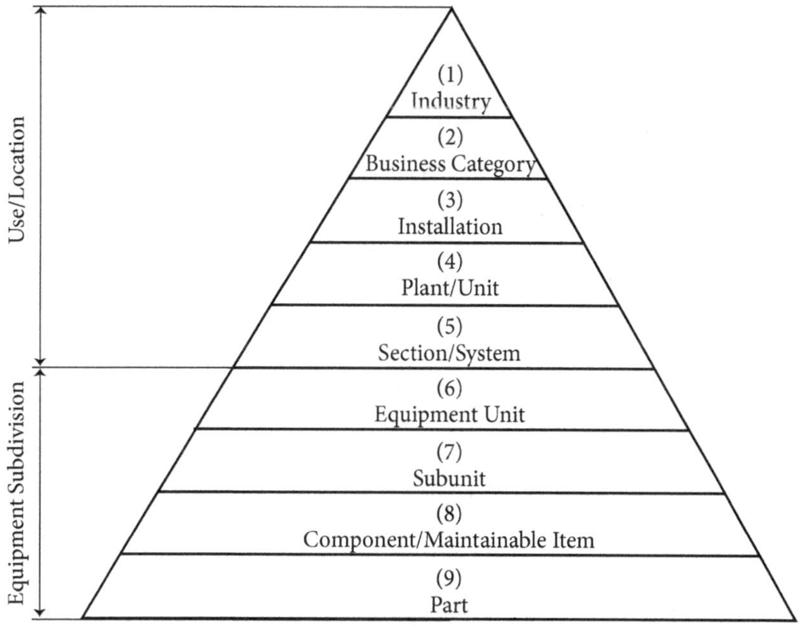

FIGURE 1-8 Taxonomy.

RULE #3—COMPLETE ITEM MASTER DATA

Let's build on Rule #1 to form small teams and create a process guide for all 32 storeroom processes and practices. Then follow those process guides and improve them through *the methodology of continuous improvement.*

Add on Rule #2—which is to form small teams and develop a comprehensive bill of materials for 98%+ of your assets. I recommend using the methods described in ISO 14224 as a guide.

And now, Rule #3, for every item classified as a spare part, you should have a complete accounting of its characteristics and important information in the component's Item Master Data table in your inventory management system.

Rule #3 is going to play a significant part in the discussions of the 32 storeroom processes and practices. The information in the Item Master Data tables is critical for the storeroom processes of New Item Setup, Order Parts, Issuing, Kitting, Parts Standardization, Obsolescence, and many more, as well as the storeroom practices of Critical Spares Evaluation,

VMI/Consignment, and Data Scrubbing. Similar to the Bill of Materials rule, this rule is non-negotiable.

For each item that you have stocked in your storeroom, there is a listing of that item and all its identifying and ordering characteristics in a computerized table that contains the item master data. As the name implies, this is the "master data" for the item. I've seen upwards of 47 individual fields that could be utilized to distinguish one part from another. Some of the characteristics that would typically be found in an item's master data are:

- Part number (your part number for the item)
- Item description
- Item location
- Where the item is used (on what asset it is used; this is from the BOM)
- Manufacturer's name
- Manufacturer's part number
- Vendor's name
- Vendor's part number
- Last purchased price
- Last purchased date
- Min/Max (or, Reorder Point and Reorder Quantity)
- ABC classification
- Unit of purchase (e.g., a case, roll)
- Unit of issue (e.g., each, per foot)

This is just a short list of some of the key information that should be documented for each item. The part description is uniquely important, as it gives the storeroom associate a fighting chance to identify the right part when requested. Consider Figure 1-9, a simple picture of two markers.

The image you see in this book is in black and white. I can tell you that the marker on the left is a pink highlighter, and the marker on the right is a black permanent marker. When describing these two completely different markers, we would be required to go out to several attributes in the description before we realize that we are discussing two different markers.

FIGURE 1-9
Two markers.

Note that both are of the brand Sharpie. That might not be clear, but they are. Suppose the storeroom had an order to create a job kit that included one Sharpie marker. That was the only instruction. What would an associate do?

Let's make it fairer. The storeroom is asked to kit a single Sharpie marker with a removable cap. Have we cleared it up? Felt tip? Quick-dry ink? With little detailed descriptors to go off, these two markers appear to be the same.

There are numerous benefits to having clear, accurate, and complete information on all the parts and materials that are stocked in the storeroom. From the short example list of information typically captured in the master data I provided earlier, a reasonable person can see the immense value of having total knowledge of each part stocked in the storeroom. It is uber-critical for the maintenance planner to have this information.

I should mention that this information is necessary and should be mandatory for all parts with a company stock number. It is common for organizations to have stock numbers for parts that they don't stock. In fact, a common term for this type of part is "order on request," or these parts are said to be stocked at 0/0 (0 min/0 max).

All items in the storeroom (parts and material) will have a stock number and will have a completely filled-out Item Master Data table. All items that have a stock number (even the 0/0 items) will have a completely filled-out Item Master Data table.

Rule #3 simply states that everything in the storeroom will be on the "system." Understandably, that is your inventory management system, which is often a feature of the computerized maintenance management system (CMMS). By "being on the system," it should be further understood that information for the parts should be accurate, up to date, and complete.

RULE #4—ACCESS

I was hoping that by the time I got down to Rule #4, the last rule, that the rules would have gotten easier to follow. But I can understand where this rule might cause the most consternation in your organization. Rule #4 creates an understanding between the storeroom and those it serves as to how it will interact with others. There are two very simple, and I don't believe unreasonable, parts to this rule:

1. A work order will be presented for every part issued from the storeroom or a part ordered through the storeroom.
2. No one should enter the storeroom other than storeroom associates or authorized individuals.

Let's break these down briefly.

I can think of 155 globally recognized and accepted maintenance and reliability metrics. That number could be bigger or smaller, but that's the number I've settled on. Of those 155, there are only two that require 100% adherence to be considered world class for that specific measure. One of those two covers spare parts issuing. All spare parts issued from the storeroom will be issued against a work order—100%.

Logic will lead you to conclude that if you can't receive a part from the storeroom without a work order, you certainly cannot order a part without a work order. 100%, without deviation.

The second part for access addresses accessibility to the storeroom. No one except storeroom personnel or authorized individuals should be in the storeroom. This is likely to cause conflict, as many who are reading this chapter only have one storeroom attendant. "What do we do during the off shifts, in case of an emergency?"

Here is a phrase that might be repeated in this book. "I can only guarantee the accuracy of my inventory to the point that I can guarantee the security of my inventory." One way to guarantee the security of the inventory is to restrict and limit access to the inventory.

Like many of you, I have a bank account. At any given time, I have money in that account. It might not be much, but it is my money. I can go to the bank and request that they give me some of my money. I can even add to the balance of my checking account by depositing more money. But at no time is the bank likely to throw open the doors of the vault and invite me in to *visit* my money. Why? Short answer: That would be dumb. A better answer: The bank must account for the accuracy of the balances, and it has controls to do that. Some of those controls involve restricting access to the money.

There are very few (emphasize *very* and *few*) reasons for a maintenance technician to be in the storeroom. After all, there is no work going on in the storeroom. Technicians can come into the storeroom if they have a work order (that's the other 100% metric) to perform a maintenance or administrative task there. Authorized individuals might include vendors, suppliers, and guests of the storeroom leadership. That's it. Access must be restricted to the bank manager and the bank tellers. I'm mixing metaphors, but you know what I mean.

The biggest offenders of both elements of Rule #4 are engineers. I don't know why. They just are. And you can take that from one.

SUMMARY

I have been consistent with these four rules for the last two decades. Believe me, if you consider these rules as a form of strategic positioning, you will truly be proactive in every sense of the definition provided by Tom Moriarty at the beginning of this chapter. You will "determine the outcomes by influencing the future."

If you adhere to these four simple, admittedly time-consuming, and sometimes confrontational rules, you will already be in the top 10% of storerooms in the country, possibly the world.

Make a strawman argument against the rules, and see if it makes sense to argue that you shouldn't have:

- **Rule #1.** Detailed and comprehensive process guides for the governance of every function of the storeroom. Process guides are created by the associates who are going to be tasked to follow them and guided by a knowledgeable expert on best practices. Everyone is trained in the processes, and the processes will be improved through continuous improvement.
- **Rule #2.** Detailed and complete bill of materials on 98%+ of your organization's assets. The BOM development process will follow an internationally established model, ISO 14224, and be accomplished by creating small teams engaged at the shop-floor level. Every single item classified as a spare part in the storeroom will appear on an asset's BOM. It isn't stored in the storeroom if it doesn't "go to" an asset.
- **Rule #3.** Every part in the storeroom has a data table listing all the distinguishing characteristics of that particular part. These data tables make up the Item Master Data in your inventory management system, often a feature of the CMMS. This rule specifically lays out the requirement that all parts classified as spare

parts in the storeroom will "be on the system." Furthermore, the Item Master Data will be complete and accurate for all stocked and non-stock items that have a stock number.

- **Rule #4.** Access into the storeroom is restricted to storeroom personnel and authorized individuals. Technicians and engineers generally have no business in the storeroom. Also, no work order . . . no part. That couldn't be any easier.

Who could argue against these four rules? Get started on these. Work on these to get them right. If you do, you will be very successful in implementing the 32 storeroom processes and practices we will discuss in this book.

WORKBOOK

If you are familiar with my earlier books, you'll know that I end each chapter with an assignment. I believe that you will get more from these types of books if you have an opportunity to consider the material you just read, discuss it with your team, and put some thoughts down. The shorter the time between having your thoughts on the subject and recording those thoughts, the more accurate your memory will be.

Here is the assignment for Chapter 1. Simply fill in Table 1-1 with a Y for yes or an N for no. If there is a question asking about a certain practice, but you don't have that practice in place at your location, answer N/A (not any). To navigate this map successfully, you would be correct in assuming that an N/A response is not ideal and that activity should be something you should consider adopting.

TABLE 1-1 Answer Y or N

Does your company have a process guide template for use?		
If so, does it contain the categories as detailed in this chapter?		
Does your storeroom have process guides or operating procedures?		
If so, do you follow them?		
If so, does your team meet periodically to improve the processes?		
Do you have a bill of materials for 98%+ of your assets?		
If not, do you currently have a small team working to build the BOM?		
Do you only stock relevant parts that actually go to something?		
Are the Item Master Data tables complete and accurate?		
If not, do you currently have a small team working to build the data?		
Is access to your storeroom restricted?		
Are the restriction policies followed? (Be honest.)		
Does your organization have the policy of "No work order, no part"?		
Does your organization actually follow that policy?		

This book offers another useful feature at the end of each chapter—a list of links to short videos on the topics discussed in each chapter.

Over the years, I have had great success reaching leaders, much like you, with my YouTube channel, Dr. John's Maintenance Minute. On this YouTube channel, I have recorded and posted many videos, including all the storeroom processes and practices. I also have these videos on my website at www.maintenanceinnovators.com.

What I discovered, quite by accident, is that some facilities are using these videos as training aids, and I'd like to encourage that practice and that exchange of ideas. One such client informed me that once a month, as part of the client's normal "toolbox talk" at the beginning of each shift, the employees break away from the routine announcements and watch a short video with their teams and discuss the video's subject as the "topic of the day."

I'll present these links as a table. In this case, it is Table 1-2. I realize it takes a bit of effort to take a link off a written page and type it into a computer, but I hope you'll appreciate the effort of trying to convey the information. Alternatively, visit my website, and look for the tab labeled "Videos." Scroll down to the storeroom videos, but also make note of the other videos available on other maintenance and reliability topics.

The only video that is relevant to Chapter 1 is on the bill of materials. But I've included some general storeroom-related videos that I believe you will find very informative and a bit entertaining. I will link to the videos on my website, as I know many of you cannot access YouTube at work. However, I will also provide the YouTube link if that would work better for you. Often, the YouTube link is a shorter link. Give this a try, and see if it is helpful. I know you will like these videos.

TABLE 1-2 Topic Links

Title	**Maintenance Innovators, Inc., Website Links**
Bill of Materials (BOM)	https://maintenanceinnovators.com/aiovg_videos/bill-of-materials-boms/
Storerooms: Service and Convenience	https://maintenanceinnovators.com/aiovg_videos/storeroom/
The Storeroom as a Core Value Driver	https://maintenanceinnovators.com/aiovg_videos/the-storeroom-as-a-core-value-driver/
Life Cycle and the Storeroom	https://maintenanceinnovators.com/aiovg_videos/life-cycle-and-the-storeroom/
Increasing Availability with an Awesome Storeroom	https://maintenanceinnovators.com/aiovg_videos/increasing-availability-with-an-awesome-storeroom/
	YouTube Dr. John's Maintenance Minute Links
Bill of Materials (BOM)	https://www.youtube.com/watch?v=L1gDXE5OYzc
Storerooms: Service and Convenience	https://www.youtube.com/results?search_query=Dr.+John%27s+Maintenance+Minute+Storeroom
The Storeroom as a Core Value Driver	https://www.youtube.com/watch?v=pAZM_GfEz38
Life Cycle and the Storeroom	https://www.youtube.com/watch?v=P0cG-YclSqc&t=10s
Increasing Availability with an Awesome Storeroom	https://www.youtube.com/watch?v=CpzFXOOuwWU

CHAPTER 2

FOUNDATION-EFFECTIVE

Even if you know where you're going,
you're still lost if you don't know where you are.

—Earl Porter

I start each chapter and have done so in previous work with a short quote that I believe sums up the essence of the chapter about to be discussed. I only draw you to this point because, as subtle as the practice is, I don't want this particular quote to go unnoticed.

Years ago, I had the good fortune to team up with maintenance and reliability legend Earl Porter for a preventive maintenance assessment at a major cheese factory in Wisconsin. I had never met Earl before, but his work is well known in the maintenance community. I was the lead auditor for this assignment, and Earl was operating as my second.

I drove up to the client site and was exiting the vehicle when Earl grabbed my right arm and essentially kept me from getting out of the car. Earl looked at me and said, "Before we go in, John, I want to ask if you know why we do assessments."

I certainly knew why we performed maintenance assessments. In fact, at that point, I had accomplished maybe a hundred assessments. To date, I'm sure that number is in the hundreds of hundreds. But here was the old man himself about to impart some wisdom to me. Typically, it would be necessary to climb to a mountaintop to get the kind of insight Earl would share. "No," I said, "I guess I really don't."

"Because," Earl started, "even if you know where you're going, you're still lost if you don't know where you are."

Picture a roadmap in your mind. Now think about a geographic location on that map, perhaps the place that you are going. Don't you need to know where you are and where you are starting from to plot your course and plan your trip to reach that destination?

Before we can start our trip, our journey, we need to establish where we are currently, and we must be honest about that location and the circumstances in which we find ourselves.

In class, I use the example of a bathroom scale. Many folks know this, and others might not. Most analog bathroom scales have a small wheel on the side or bottom that can be used to reset the scale to 0 in the event it does not reset to zero without weight on the scale. Now, it is possible and has been the practice of more than a few people to dial that scale back about five pounds to compensate for an otherwise bad day. That's not being honest, nor will it make your clothes fit any better.

Stand on that scale, own that value, and make a plan to journey forward to a lower weight and better health.

Before you can start any get-well or get-better program, you must establish where you are for better or worse. And those on the journey with you will also need to understand and agree on that starting point.

In Chapter 1, in the Workbook section, I asked you to answer yes or no to some simple questions about the rules that were established in that chapter. There is almost a 100% probability that some of you answered yes by thinking, "Yeah, we've got something like that." Or "We're doing pretty well with that."

Aren't you metaphorically dialing the scale back five pounds with those answers?

We are about to get into the *meat* of this book. In order for you to move along the continuous improvement continuum with any velocity at all, you must first own where you are.

WHERE TO START?

I'm going to detour slightly from the initial creation of this roadmap with my colleague Stephen Miller. When Stephen asked if there was an order in which organizations should approach the storeroom regarding creation or improvement, the answer was yes. Yes, there is. Over the years since Stephen and I had that conversation, I've come to realize that not many organizations are *creating* a storeroom; rather, they are *improving* an existing storeroom. As such, those organizations are likely to have many of the processes and practices already in play. Albeit, not formally documented or consistently executed. The roadmap still applies, even for mature storerooms. The processes and practices are usually always being executed. Even the smallest storerooms have many of the 32 at work.

With that in mind, I also want to stress that this first zone, the Foundation-Effective zone, is unapologetically the most crucial. As the most crucial, I would insist that organizations ensure that these elements are 100% enforced, engaged, and active before traveling to any other zone. If you have a mature operation already, go back to this first zone and make sure you have all these elements right. Don't mess with that scale!

Let's take a close look at Figure 2-1, which shows the Foundation-Effective zone.

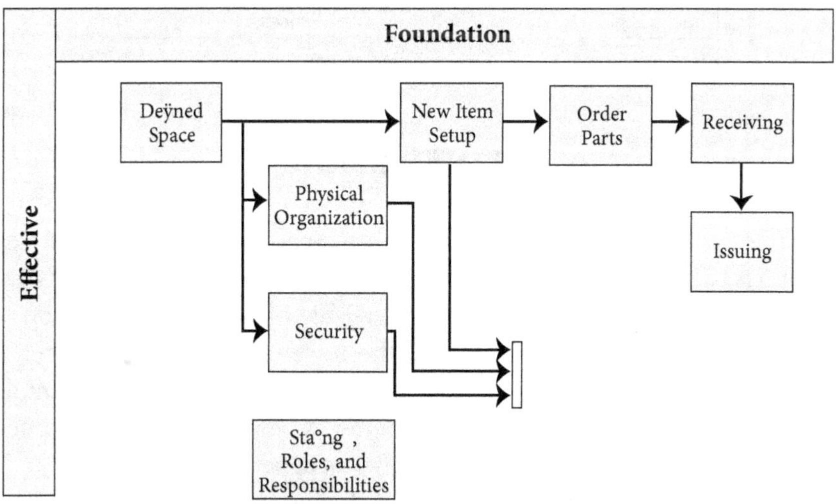

FIGURE 2-1 Foundation-Effective zone.

This zone has eight elements in it, and all are critically important for the journey ahead. If you'll allow me to make an analogy of a roadmap used for a family trip, this is the portion of the map where you are setting up everything you'll need to have a safe and enjoyable journey. When working with clients to build or improve their storerooms, we always start right here. Unapologetically.

For the remainder of this book, in these chapters pertaining to the different zones, I'm going to show a table of the elements and in which category they are typically placed (e.g., process, practice, organizational, or KPI). Your operation may be different, but the principal use of the map will be the same. Consider this table as the legend printed on a regular map to help viewers better understand what the symbols and marks indicate. Table 2-1 indicates what category each element corresponds with.

TABLE 2-1 Legend for the Foundation-Effective Zone

Element	**Activity**
Defined Space	Organizational
Security	Organizational
Physical Organization	Organizational
Staffing Roles and Responsibilities	Organizational
New Item Setup	Process
Order Parts	Process
Receiving	Process
Issuing	Process

DEFINED SPACE

As Table 2-1 indicates, the element of Defined Space is an organizational element. There is a bit of a chicken vs. egg controversy that sometimes brews with clients on where to start in this Foundation-Effective zone. Let me explain.

First, it is important to note that the rows on the roadmap (labeled "Efficient" and "Effective") are in the order they are in because you want to get established as being effective before improving your efficiency. If the order were reversed, you could become very efficient at being very ineffective.

I see this with companies that want desperately to adopt bar coding before they've even mastered the process of issuing parts. I tell my students they'll struggle with the bar coding principles if they haven't first mastered pencil and paper. Give that some serious thought. Become *effective* and then improve your *efficiency.*

The chicken vs. egg debate centers on which of the eight elements in this first zone to start with. I'll welcome any dissenting opinions on this, but my experience and practice lead me to advise starting with the Defined Space element. This is where we begin.

"Defined Space" is simply the name given to the process of declaring what space is reserved for the storeroom and storeroom activities. Where does the storeroom have jurisdiction, total and absolute authority? This declaration must be in writing or documented in some official manner. It can simply be a facility map that clearly indicates that certain spaces are set aside for the storeroom.

The justification for starting with this element is actually quite clever. Right at the start, we want to announce that "this *is* the storeroom" and "this *is not* the storeroom." From this moment on, when you say "storeroom," you are referring to *this* space.

For example, take Rule #4, "No one should enter the storeroom other than storeroom associates or authorized individuals." That means in *this* space.

Or consider Rule #3, "Everything in the storeroom will be on the 'system.'" That means in *this* space.

There is a conversation that I've had with many maintenance managers and plant engineers, and it is often a very soul-searching point. You (maintenance manager, engineer) have hired a person to run the storeroom. Now stay out of that person's way. I say this as a former maintenance manager and plant engineer. I've made those mistakes and committed those micromanagement crimes. So much so, in fact, that I've covered it for all of us. There is no need for you to do it as well.

The processes and practices that we will spell out in the remainder of this book apply to *this* space. Doesn't it make sense that everyone agrees on what *this* space is? Does it further sound reasonable that within *this* space, the storeroom leadership calls the shots and runs the operation? It certainly makes sense to me.

Document in a written governance policy or show on a facility map clearly what spaces the storeroom comprises. This could include any or all of, but not limited to, the following:

- Main storeroom
- Satellite locations (lockers, rooms, drawer bins, etc.)
- Off-site storage locations
- Vending machines
- Receiving quarantine locations
- Shipping holding areas
- Bench stock materials in the maintenance shop or production area
- Job kit staging locations

The proclamation is simple: "Let all who hear this understand that the policies, procedures, and rules governing the storeroom and the storeroom operations will be applicable within all these geographical locations. The storeroom is in charge here, and you are not."

Talk about putting the weight on the storeroom's shoulders. You've just announced that you alone are in charge of these locations. You are responsible. Now start acting like it.

Once the storeroom's defined space is clearly established, documented, communicated, and understood, the real work begins. The space should be functionally operational. Several of the other 31 processes and practices will speak to the layout and functional aspects of the storeroom, but it starts here with defining the space, setting up the space for functional operation and success, and communicating to those who utilize the storeroom how the space is set up.

Figure 2-2 shows an example of a storeroom layout. This particular layout is for a Midwest pharmaceutical company and is an example of a Zoned-Hybrid approach. The company's main storeroom occupied a stand-alone building, which is always an envious arrangement.

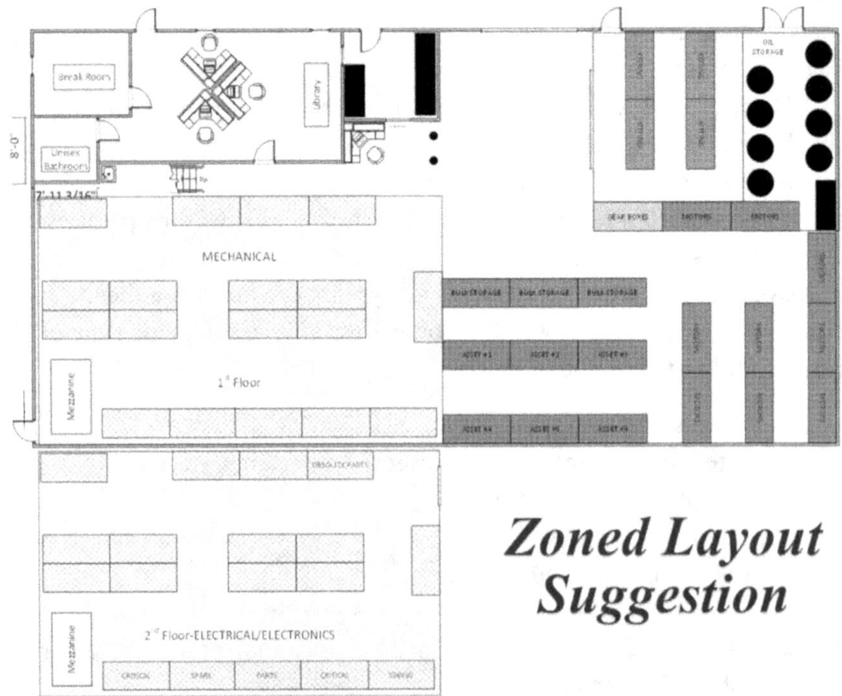

FIGURE 2-2 Storeroom layout.

There are many key features of a storeroom layout that are considered best practices. And as much as anyone is able to influence these features being present, the effort must be made. Key to the list of storeroom features are:

- Roll-top door to the outside so goods can be received directly in the storeroom
- Open space for receiving materials and building kits
- Enclosed office space for completing administrative functions
- Separate job kit storage area
- Adequate shelving (with OSHA-recognized spacing)
- Adequate bin storage
- Enclosed office space with room for a technical library (with a copier machine)
- Independent lubrication storage with design enhancements (e.g., no floor drain)
- OSHA-recognized lighting levels
- Separate and more secure storage for critical spare parts
- Separate storage for obsolete and quarantined spare parts
- A space for processing items to be shipped out (e.g., repair, return to supplier)
- Environmentally controlled storage for electronics and sensitive components

Although Figure 2-2 shows a storeroom layout with a mezzanine over part of the space, I don't recommend mezzanines. It never made sense to me to lift something to an elevated platform and then have to lower it back down again. It is a sure bet that if you have a mezzanine in your storeroom, it is because you ran out of space. Very rarely does someone envision a mezzanine storage area and think, "That is the most logical idea I've ever had." In the case of Figure 2-2, I simply couldn't talk this client out of having a mezzanine.

Storerooms should be visual workspaces. Locations should be clearly marked, and standardization should be the order of the day. Standardization is sometimes misunderstood. In the case of standardizing a visual workspace, know that rows/aisles should be numbered or lettered (1, 2, 3 or A, B, C), and the shelving units on that row or aisle should be the opposite (A, B, C or 1, 2, 3).

For example, Row 1 is made up of shelving units A, B, C, and D. The shelves on shelving unit A are numbered 1, 2, 3, etc., starting from the bottom. Very important—start from the *bottom*. The locations on shelf 1 are A, B, and C (left to right). Alternating between numbers, letters, numbers, letters, and so on. The opposite works just as well—letters, numbers, letters, numbers.

Shelf capacity should be marked for weight restrictions. A lot of these details are folded into the housekeeping/organizational methodology of 5S (sort, set in order, shine, standardize, and sustain). Figure 2-3 is an example of a good, visual workspace indicating that you are at Row A.

FIGURE 2-3 Visual workspace.

SECURITY

I don't think anyone will agree with me, but I almost feel as though a storeroom does not need to be locked down. Seriously, hear me out.

I am a veteran of the United States Air Force. When I served, I was a flight line aircraft maintenance officer. My job was on the flight line, with the crew chiefs and sergeants who maintained the aircraft. I will concede that the Air Force base had a fence around it and was patrolled by armed military police. However, the flight line, where the aircraft were, did not have such a barrier. No fence. Not where I was. What the flight line, or the tarmac, did have was a red line painted around the perimeter. You only crossed onto the tarmac at a designated and controlled opening. You never crossed the "red" line. The penalty for violating this very simple rule was being dropped, face down, with a gun pointed at you. An extreme example? Yes, of course. I wanted to illustrate that people avoid doing foolish things when they are trained properly and held accountable to follow their training. Unfortunately, that isn't generally the case in industry.

In my consulting experience, I've noticed a similar environment in most places where I've worked. Usually, when I enter a facility, I go through the requisite training videos, sign a form or two, and am issued a visitor's badge. As I work in the plant or facility and get to know some of the folks, I find that I'm given "a little extra leash." I don't do this in practice, but I feel that I can wander the facility almost unchallenged.

I can virtually walk around a company's principal supplies and finished goods without even being challenged. But to access the parts room, I need to wait for someone to open the door. What do you suppose the inventory value is of the storeroom compared with all the supplies it takes to make your product and all your finished goods?

Why don't people mess with the production inventory? Why, when given a chance to goof with spare parts, they almost can't help themselves?

Why do people only cross onto the tarmac at the designated access point?

Because they know better. I'm just puzzled why the storeroom can't be like that.

OK, I told you that you wouldn't agree with me. That's just my position on the matter of storeroom security. I temper that personal belief with the real-life understanding that I can only guarantee the accuracy of my storeroom inventory to the point that I can guarantee the security of my storeroom inventory. That realization and the reality of it all has prevented me from recommending the foolhardy approach of having no security.

It is exactly like a bank. The bank manager can guarantee the accuracy of the inventory of the money to the point of guaranteeing the security of the money. Rule #4 (remember that one?). Absolutely restrict access to the vault . . . mixing metaphors (again).

Typical storeroom inventories can be valued in the multimillion-dollar range for even small to moderate-size operations. For that reason, the storeroom should be locked down. Those parts are company capital, just like the money in the bank is company capital. Most people would not leave $2 million in a room with an open door.

Lock it down. With today's technology, there is no excuse for not having digitized, badged entry. Only storeroom associates or authorized individuals should be in the storeroom. Authorized individuals are those who are invited into the storeroom. Only salaried supervisors or hourly lead men should have access to storerooms that do not have personnel manning the inventory over the weekends or off shifts.

There is no reason for a maintenance technician to be in a storeroom unless that technician is working (with a work order) on a job. If a part is needed during the shift, the supervisor gets the part. That's what a supervisor does.

Our challenge now becomes to influence the future so that no one outside of storeroom personnel ever has a need to enter the storeroom.

PHYSICAL ORGANIZATION

There must be someone in charge. That is what this organizational element is addressing. The Society for Maintenance & Reliability Professionals alludes to the importance of having a clear authority in the storeroom when clarifying the metric for stores transactions in its *SMRP Best Practices*: "This metric should not be applied to open storerooms where the storeroom clerk position is filled by multiple employees, who at various times are performing all the different responsibilities of a storeroom clerk. By definition, the storeroom clerk position is an assigned employee" (p. 275).

SMRP is warning those using its best practices for reference that, for this metric, they cannot just have anyone *floating* in and out of the storeroom performing storeroom functions. As the last sentence in the previous paragraph clearly states, the clerk is "assigned" to the storeroom. There must be someone responsible for the storeroom operation. This isn't a part-time gig.

Let's give some really deep thought to the staffing levels that might be necessary to run what we hope is an effective and efficient storeroom. We do know that regardless of how many folks are assigned to work in the storeroom, they will be executing 32 processes and practices—at a minimum. Regardless if there is one person or seven . . . there are still 32 functions.

It will take dedicated and focused professionals to secure and grow a high-performing storeroom. Table 2-2 shows some values, from *SMRP Best Practices*, that we know will need significant attention to reach the best-in-class values (pp. 257–283).

TABLE 2-2 Storeroom Best Practices

Metric	**World-Class Target**
Stores inventory turns	> 1.0
Stockouts	< 2%
Inactive stock	< 1%
Transactions (per associate)	100–140 per day
Storeroom records	5,000 per associate
Maintenance material cost	</= 50% of budget

For reasons that escape this author's ability to understand, there are still organizations today in the United States that insist on only having one resource assigned to the storeroom. Again, that one person is responsible for the faithful execution of at least 32 functions. How many functions can the average person complete, almost simultaneously, with superior results? That was a rhetorical question. An organization is not likely to consistently perform at the peak levels described in Table 2-2 with one person in the storeroom. An organization certainly will not even get to those peak levels with only one person assigned and working in the storeroom.

Where it is true that to comply with the Physical Organization element of this first zone, you only need to have one person assigned, how many people are required to staff an effective storeroom? Even SMRP, in its work, hesitates to attack this topic head-on, except to caution that in the best storerooms, one attendant can manage 5,000 line items. We are left to figure out what would be a reasonable level of responsibility for a storeroom operator in an average or below-average storeroom operation. Some anecdotal sources indicate that an average attendant in an average storeroom can manage around 2,500 MRO (maintenance, repair, and operations) items. The difference between the best and the average storerooms is the presence and use of mature processes and systems that run effectively and efficiently. A high-performing storeroom has all the systems in place to virtually run itself. All activities would need to be executed consistently at the highest level. That is a lot of ground to cover, and most storerooms are not likely to grow from average to excellent with one person. I've never seen it, and neither have you.

But take a breath. We don't automatically throw people in (or at) the storeroom thinking that the problem will be fixed. More people with no direction is never the answer.

In physics, a vector has speed *and* direction. Sometimes, in our haste to "get better quicker," we end up adding people into a bad system and end up frustrating good people. We become all speed and no direction.

One of the first steps to take in the Physical Organization element is to draw out what you want your organization to actually look like, and therefore *perform* like.

Figure 2-4 is an example of an organizational structure. This is the organizational structure I built at the first plant I ever worked.

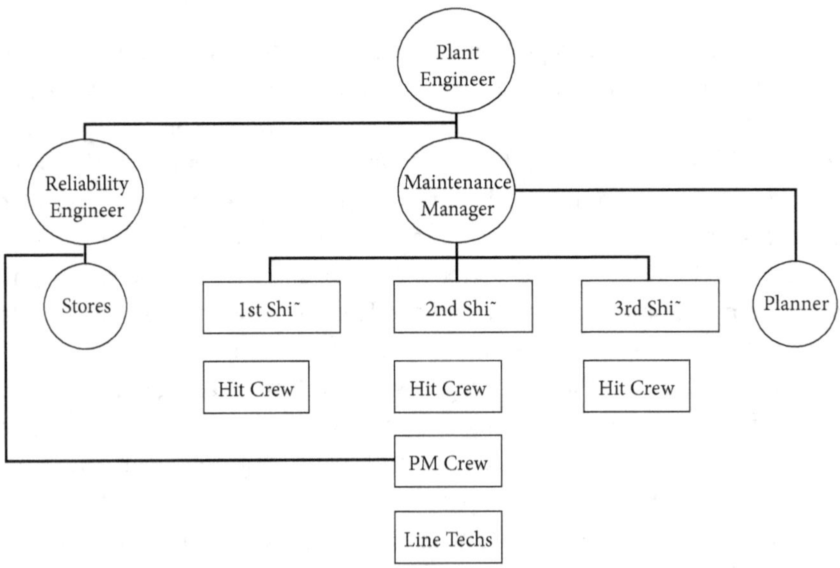

FIGURE 2-4 Maintenance organizational structure.

My maintenance organization didn't look like this when I started. With the help of an excellent maintenance consultant, I drafted this image and told my boss that this was what I hoped to develop while I was building up the maintenance department. My boss, who was a great supervisor, challenged me on the number of people I wanted to add and worked to make sure I believed in my own idea so he could defend it. I actually built this . . . in a very militant union company. I'm not going to kid you. It was a fight, but we got it done.

I was the plant engineer. Notice that my storeroom supervisor reported to the reliability engineer. The dedicated PM crew also reported to the reliability engineer. That structure suggestion was courtesy of the consultant I engaged.

I'm still thankful for the consultant I was working with for planting this little organizational nugget. It was a great alignment.

What does your organization look like; more specifically, whom does your storeroom leadership report to? It is not only important that you make sure you satisfy this element by having someone in charge of the storeroom and the storeroom functions. But also that the storeroom leader reports up the chain to someone who is committed to the mission of the storeroom and preserving the inherent reliability of the assets and the parts stocked in support of that asset.

The storeroom organizational structure itself needs to reflect the importance of the storeroom and, specifically, the functions that your organization feels the greater commitment to.

Figure 2-5 is an example of what a storeroom organizational structure might look like. Don't lose your mind over how many positions are shown on this organizational chart. Instead, ask yourself if this arrangement lends itself to supporting the fundamental elements of a very high-performing storeroom. And doing so consistently.

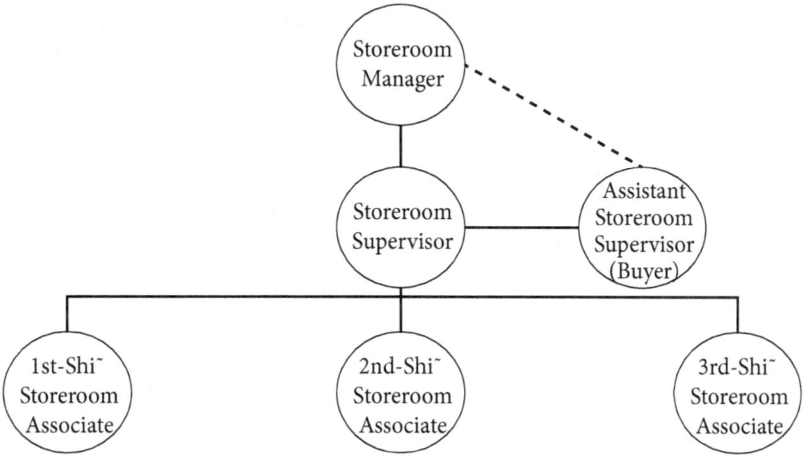

FIGURE 2-5 Storeroom organizational structure.

I've written about this before. Roger Lee challenged us to take a hard look at our organizational structure and understand that what got us "here" may not get us "there" in terms of improvement: "A proper team structure is required to maximize your efforts when implementing a cultural change to any organization. You are just feeding your maintenance insanity when you leave the same structure in place and the same people in the same roles forever" (p. 134).

Draw out what your storeroom's physical organization should look like, and always keep that vision out in front. Speak it daily. Speak it into existence.

The structure doesn't begin to take its final shape, nor are we able to *speak* to the depths of its merits, until we can concisely, yet conclusively, spell out the staffing roles and responsibilities. This is our next element.

STAFFING ROLES AND RESPONSIBILITIES

It is vexing and has been a decades-long observation by this author as to why organizations refuse to document roles and responsibilities. If not an outright refusal, then an insistence on getting them wrong. People often mistake "roles," "responsibilities," and "job descriptions" as synonyms. They are not.

The rationale for documenting roles and responsibilities is simply an argument that it makes sense for associates to know what roles they play in the execution of organizational functions, and the responsibilities they have within those particular roles. In that way, the associate can be aware of what skills, talents, and abilities are needed, request any training necessary, and secure the tools required to comply with the task at hand. Supervisors will know what training, tools, and direction to provide.

There are also directives encouraging us to have documented (written), detailed roles and responsibilities for our associates.

The first pillar in the *SMRP Guide to the Maintenance and Reliability Body of Knowledge* is Business and Management. This is a perfect place to establish the unwavering requirement for documented roles and responsibilities. In that section, SMRP unambiguously informs us that we will be "providing written definitions of roles and responsibilities" (p. 4).

Clear-cut directives also appear in ISO 55002: "Persons doing work under the organization's control understand their roles and responsibilities" (p. 2).

Under the section for Organizational Roles, Responsibilities and Authorities, ISO 55002 states: "The responsibilities and authorities of key functions should be defined. This should include both internal and outsourced roles and responsibilities" (p. 7).

The following simple example should clarify the differences among roles, responsibilities, and job descriptions. Hopefully, you will gain an appreciation for having all three and having them documented and communicated.

If you look back at Figure 2-5, which shows the storeroom organizational structure, you will see that there is a functional title for a storeroom associate. A storeroom associate has several roles. Table 2-3 is an example of some of those roles and some corresponding responsibilities.

TABLE 2-3 Roles and Responsibilities

Role	**Responsibilities**
Inventory clerk	Receive items
	Inspect items
	Shelf items
	Cycle count items
	Issue items
Shipping clerk	Request RMA
	Package items
	Apply proper postage
	Complete paperwork

The following might be an effective job description for a storeroom associate:

Position

Storeroom associate

Job Description

Long-term team player constantly engaging in challenging, permanent work in an often chaotic environment.

Associates must possess excellent communication and organizational skills and be willing to work variable hours when needed, which will include evenings and weekends (on call).

The storeroom associate is responsible for providing service and convenience in fulfilling any stocked item request.

The storeroom associate is part of a dynamic team and is required to be part of the continuous improvement effort to improve 32 storeroom processes and practices for consistently superior service and convenience.

Responsibilities
Receive delivered items, both physically and through the inventory management system

Document "goods movements" in the inventory management system

Housekeeping in compliance with the established 5S program

Cycle counting in accordance with the established program

Pick and kit parts as directed by the weekly maintenance schedule

Issue spare parts and other stocked items as requested (but only if the issue is against a valid work order)

Run usage and performance reports, track and post metrics

Adjust inventory count within the inventory management system

Perform preventive maintenance tasks on stored components, document those PMs in the inventory management system

Answer the phones, respond to email messages, attend meetings as required

Drive forklifts and operate manual and powered pallet jacks

Create labels, ship (UPS, FedEx) packages and components to be refurbished

Possibility for Advancement and Promotion
Storeroom associates are brought in under an entry-level ranking; advancements are commensurate with a rank increase and a 2% raise at each incremental level until level 9 is reached. This

is based on performance and compliance with the training initiatives spelled out in the onboarding material.

Advancement and promotion to any nonskilled position in the storeroom and/or the maintenance department is possible, based on performance and attitude.

Previous Experience
None required

Wages and Compensation
As published

Benefits
The storeroom associate is an hourly position. Individuals are compensated for wages based on seniority with the company. Benefits are commensurate with the information posted in the *Employee Handbook.*

How many employees can perform their jobs consistently at the highest level without knowing what their jobs are? By clearly defining roles and responsibilities, you give your team a fighting chance to "get right" those activities that you deem essential to the success of the storeroom's effort. As I hope you can see from the brief example, job descriptions are another product altogether and much different than roles and responsibilities.

Before we leave this element, there is just one more point to make. Look back at Chapter 1 where a sample process guide was provided as an example for you to use in developing your guides. In that sample is a RACI chart. You will see the letter R written in the columns on that chart. Intersect this column point with the row's activity. For the sample process guide for Issuing, that functional title is responsible for that activity at this intersecting point. This is a great method for populating the responsibilities portion of roles and responsibilities.

NEW ITEM SETUP

This is the first process we reach in the Foundation-Effective zone. As a gentle reminder, I strongly recommend that you don't attempt to establish any of the processes or practices elements until you've completely addressed and consistently executed the organizational elements we just discussed at the beginning of this chapter.

As the element title suggests, this process is where we establish the criteria for putting components in stock. This same process is used when we don't want to stock the item, but we'd still like to create a company part number for it. For items that aren't stocked yet have a stock number, they are referred to as "0/0 (0 min/0 max)" or "order on request."

The New Item Setup process should be used for any item requested to be stocked or with any item that will have a company stock number. Even nonmaintenance spares (e.g., safety, operating supplies, office supplies, etc.) should be handled like spare parts for storeroom setup.

The rationale for a formal process is quite simple, and I would argue that we all have a horror story we could share. Without a formal process for establishing an item in inventory, the storeroom associate is likely to come to work one day and find an empty box or a plastic wrapper on their desk with a note that says, "Stock 10 of these." It is also just as likely that the associate will find an actual spare part, or worse, a used-up old part sitting on the desk with a similar note.

Maintenance departments have sent their spring-cleaning "finds" from a maintenance shop 5S event over to the storeroom on pallets with a request to stock the parts, as have engineers with their project work residue. We should stop this nonsense. Immediately.

The New Item Setup process is also often referred to as the "Add to Stock process." In all instances, this is the process to follow to have an item established as a stocked item or have a stock number created for it.

Your New Item Setup process should communicate some nonnegotiable requirements:

- All requests for stock additions *will be* in writing, using a detailed form.
- The form used will have a section for the requester to complete and for the storeroom and the purchasing department to complete as well.
- The requester is the starting point for the form. Questions the requester must answer might include:
  - What is that part?
  - What's it called?
  - Where does it go?

  - Where did you get that one?
  - How much did it cost?
  - What does the supplier call that part?
  - How many should we stock?
- The requester is obligated to make sure that the part doesn't already exist as a stocked item, or as a documented item with a stock number (0/0 stock).
- The requester's supervisor will confirm that the part is needed in stock, or that it should have a stock number.

On a workflow diagram, these steps would look like those shown in Figure 2-6.

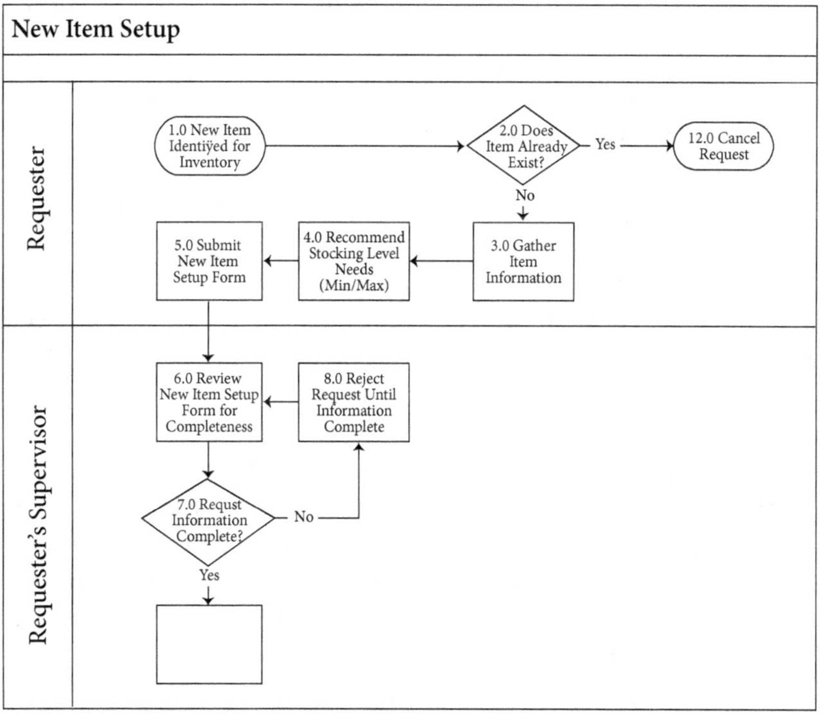

FIGURE 2-6 New Item Setup workflow (partial).

To continue with our list of nonnegotiable requirements:

- The maintenance planner will be the only person sending/giving the form to the storeroom (the planner is the liaison to the storeroom).

- No items will be stocked in the storeroom (or even located in the storeroom) until they are officially "set up" in the inventory management system.

There are other features that I believe can take the New Item Setup process to another level:

- The New Item Setup form mirrors the Item Master Data fields, exactly (discussed in Chapter 1)—imagine how easy it would be to enter new items into your inventory management system if you have all the information and in the proper order.
- The form could also be used for:
  - Identifying suitable substitute spare parts
  - Confirming items to remove from inventory, for example, those confirmed to be obsolete

Figure 2-7 shows an example of a New Item Setup form.

The sample New Item Setup form shown in Figure 2-7 does not have all the elements I recommend, but it has several. Use these tips as the catalyst for creating your documented process for stocking new items in the storeroom.

Against my better judgment, I'd like to wade into the pool of discussion on part numbers and part numbering. The importance of this particular topic, over almost any of the other fields in the Item Master Data table, is vital to understand. The part number is how everyone is going to communicate with everyone else. Let me explain.

At its simplest, the technicians will tell the storeroom associate what part they need by communicating a part number. By recording the part number, the planner will request that a part be kitted for a planned job on the pick list. Technicians who are required to enter the storeroom to retrieve their parts must document the part number along with the count of the parts they are taking.

When an organization wants to add an item to its inventory, the organization is required to use a New Item Setup form, as discussed. And as noted, that form has individual sections for the requester, the storeroom, and procurement to complete. Procurement must coordinate with your company's headquarters. Procurement is checking to determine if the item already exists on the company register. That item search will undoubtedly be by part description, but will ultimately result in a part number if the item already exists in your corporation's registry.

New Item Setup

Requester:		Date:	
Reliability Engineer Approval:		Project Name:	
Maint. Manager Approval >$10,000		Manufacturer's Part #:	
Part Description:			
Manufacturer:		Drawing #:	
Supplier (where purchased):		Price:	
Replaces existing part? #:		PO#:	
Minimum Qty/Reorder point:		Est. Normal Lead Time:	
Max Stock Level:		Any Shelf Life Restrictions? :	
Repairable(vendor or inhouse):		Impact: (A=Plant Down B= Line Down C= Machine Down D=Little/No Impact)	
Stock or Buy item:			

BOM Line - Area - Machine - Func Loc - Equip Number & assembly (ALL THAT APPLY)

STOCKROOM USE ONLY

			Initials Date
Part #:		Created in system	
Bin Location:		Photo taken & attached	
Purchase UOM:		Area Planner - BOM	
Supplier #:		Requester Notified	

FIGURE 2-7 New Item Setup form (client example).

There are generally two formats for a part number:

- Intelligent
- Nonintelligent

Arena Solutions offers a great description for each:

> **Intelligent**—Also referred to as "significant." The part number contains descriptive and informative details that provide significant information about the part. With this type of scheme, a part number generated for a resistor might be RES-100-0003, where "RES" stands for resistor, "100" is the resistance value in ohms, and "0003" is a serialized suffix.
>
> **Nonintelligent**—Also referred to as "nonsignificant." The part number does not provide any information about the part. Nonsignificant part numbers are typically serial (pulled in numerical order), regardless of the type of part. By using this part numbering system, a resistor could be assigned part number 000416782.

If your organization does not already have a preferable format, I strongly encourage the nonintelligent format. It is less complicated and lends itself better for consolidating stock and merging storerooms (e.g., acquisitions). You will find yourself working hard to resist the temptation to use intelligent part numbers. Avoid it if you can. The temptation is strong, but the horror of subjectivity that is going to plague your future is almost certain. Will a bearing be identified as "bearing," "brng," "bng," or simply "B"? I've even seen "baring" once.

The nonintelligent number is simply the very next number generated. No subjectivity. There is no confusion if a hydraulic motor is a pump or a motor or if a reducer is a gearbox. Very simple format. I suggest you use it.

It makes sense to have a process for setting up stocked items before ordering items to fill the stocked inventory.

ORDER PARTS

There are only a few reasons that parts are ordered—aside from parts that are ordered for a capital project or a major shutdown/turnaround. Routine spare parts are ordered specifically for three reasons:

1. As initial stock
2. As replenishment of stock
3. For a planned maintenance job (against a work order)

In all instances, the parts will be purchased from an approved vendor. If not, and most organizations have experience with this, a P-card, or purchase card, might be used. Sometimes the P-card is used even if the supplier is an approved vendor. This is often done for expediency.

Initial stock is the ordering activity that is used when a new item has been created in inventory based on a requester's input. From the previous element discussed, it should be clear that an item is not to be brought into the storeroom until it is set up in the inventory management system. The maintenance organization will often purchase two parts for a maintenance assignment when only one part is needed for a repair. Maintenance has the intent of putting the extra purchased item into inventory. This is not a practice any reasonable and seasoned maintenance manager should promote. However, as far as the storeroom is concerned, the folks there should not be concerned by this practice. The surplus part will be put into inventory only after the requester fills out the New Item Setup form and after the item is set up in the inventory management system. The storeroom personnel don't care where the maintenance department stores that extra part in the meantime; it just isn't going to be in the storeroom until the paperwork is done. (*Author's note:* This is a critical test point of the storeroom's authority to deliver best practices by refusing to engage in less-than-exceptional storeroom practices.)

Replenishment stock is the most common spare part purchasing activity. The idea is to replenish inventory to a level that lies between a point of two extremes: too much and not enough. Even though most maintenance organizations refer to stock limits as "min" and "max," most computerized maintenance management systems have changed to using "reorder point" and "reorder quantity."

When a part is issued, the item is subtracted from the inventory. This action is often referred to as a "goods movement." From time to time, those issues are going to hit a reorder point. The inventory management system will be triggered by this occurrence and generate a suggested reorder quantity for that part. The quantity on the reorder will be the reorder quantity. Not all lesser or homegrown CMMSs have this MRP element. MRP is an acronym for materials requirements planning. It has also been referred to as "materials resource planning." This MRP is the subroutine of the inventory management system that checks for these triggers and then alerts when the event occurs. The reorder point and reorder quantity are some of the information documented in the Item Master Data table.

The reorder point and reorder quantity should be as predictable and as smooth as the reordering model (Figure 2-8) used in the American Reliability & Maintenance Institute's Certified Storeroom Manager/ Technician course.

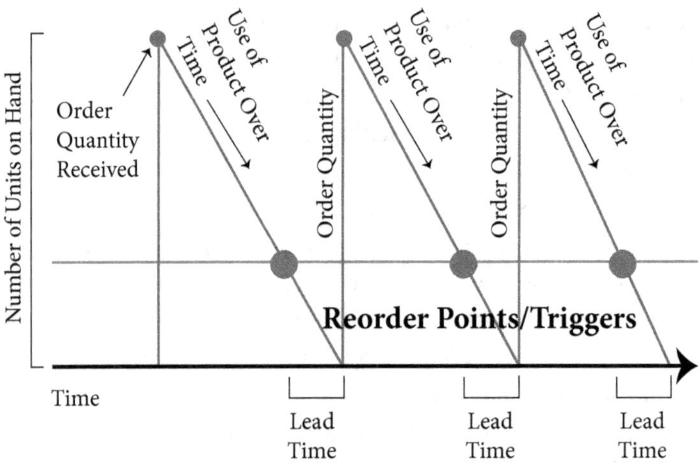

FIGURE 2-8 Reordering model.

The most significant step in the process of replenishing stock is to confirm that the stock actually needs replenishing. It is not unusual, especially if there are issues with inventory accuracy, for the computer accounting to be misaligned with the physical count. In short, the remedy is to go and look. This little, almost insignificant step of confirming the need for replenishment is often what separates good storerooms from great storerooms.

Your Order Parts process guide should reflect this confirmation step of ensuring that parts identified as replenishment candidates are truly at a level that reflects the need. Consider Figure 2-9 as an example to use when creating your Order Parts workflow.

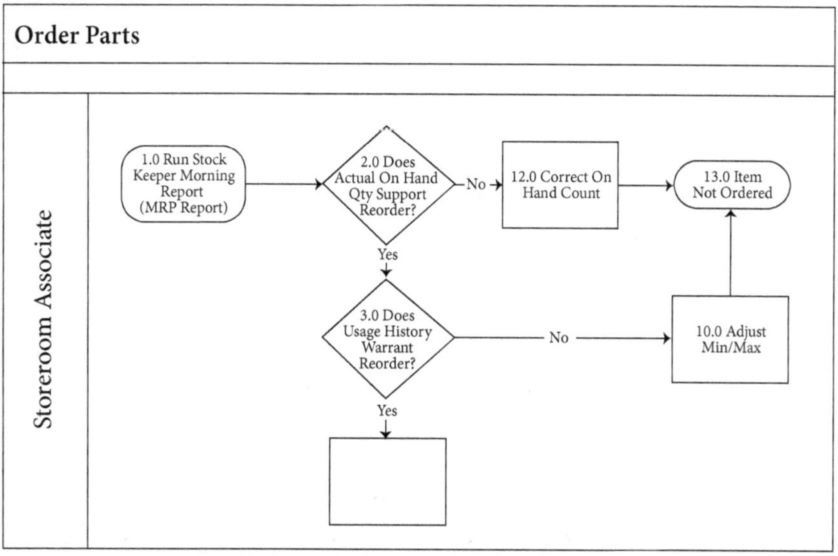

FIGURE 2-9 Confirming stock levels.

I want to make this next point unambiguously clear. A part will not be issued for a maintenance job without a work order, nor will a part be ordered for a maintenance job without a work order. All spare parts ordered for a maintenance job should be ordered through the CMMS work order. Most major CMMS software programs have a materials or components tab. Ordering through the work order is a best practice.

Ordering parts directly for all planned maintenance jobs instead of pulling from stocked inventory should be the vision of every maintenance organization. True high-performing and world-class maintenance organizations should have the aim of keeping inventories very low. According to the SMRP, turn ratios for best in class should be greater than 1 (*SMRP Best Practices*, p. 257), with anecdotal evidence suggesting that the number should be around 1.4. The turn ratio will be discussed and explained later in the book.

A storeroom that can operate with a turn ratio of 1.4 consistently is a storeroom that is partnered very closely with an exceptional preventive and predictive maintenance program and the world's best planning and scheduling efforts. A storeroom with a turn ratio of 1.4 is a storeroom that doesn't stock a lot of inventory. Maybe a few belts and a few bearings. Unlike most storerooms that have hundreds of belts and hundreds of bearings.

To this day, it is still a common practice for people to order parts and other job materials through the storeroom. Once the part is received, the storeroom will dutifully "receive it in," tie a red tag to it, place it on a designated shelf, and call the engineer, supervisor, or planner to "come and get their parts." Those parts are likely still sitting on that designated shelf to this day.

By ordering these job essential parts through the work order, as previously mentioned, when the part is reconciled with the purchase order when received, the work order number will be present on the P.O. This alerts the receiving clerk that the part was ordered for a specific job, and allows the clerk to put the part directly into a job box for a formally planned and scheduled job that is being kitted. Remember, the part was ordered with a work order number. Kitting is a process that will be detailed later in this book.

I've mentioned more than a few times in this book about the practice of creating a part number for an item that will not be physically stocked. It is a good practice to create the stock number and require that the Item Master Data table be complete for that item to fulfill an ordering requirement sometime in the future. That is where we find ourselves right now. For the instances where a part needs to be ordered, that part can now be ordered (through the work order) by referencing its part number instead of a long string of component attributes. This difference is a choice between ordering a W00783451 or a Thrust Ball Bearing 17 × 30 × 9mm DIN 711 Flat Resting Surface W236C. In which ordering number are more transposition errors likely to occur?

The Order Parts process is naturally followed by the Receiving process.

RECEIVING

I'm often asked, "What's the best storeroom you have ever been in?" I can tell you with certainty that it wasn't a storeroom I was responsible for. No, indeed. The best storeroom, in terms of operation, that I've ever seen was at a chemical plant in South Texas. It wasn't much to look at, but the lady who ran that storeroom was process-driven. If the process said to do it one way, then that's the way it was. Processes were corrected and routinely improved to capitalize on new best practices.

The *thing* that stood out about this storeroom, aside from its almost 100% kitting operation (there were no parts issued at the storeroom window—and there were never any emergencies) was its receiving process.

I hesitate to describe the process, but I'm going to do it anyway. Caution should be taken to *not* copy this process because you don't likely have the wherewithal to pull it off. The storeroom manager at this facility was exceptional in almost every conceivable way, and her team worked like a well-oiled machine.

In the Foundation-Effective zone organizational element of Defined Space, I mentioned that having a roll-top door to the outside was a wonderful feature so the storeroom could receive deliveries directly. This storeroom in Texas had such a door, but it wasn't as part of a loading dock, which would have been ideal; it was instead at ground level.

As I mentioned, the storeroom manager was very process-driven. She understood that in order to "sell" a part from the storeroom, it first had to be received. Receiving a part isn't just the process of taking it from a delivery driver. It includes the reconciliation of the part against a purchase order and confirmation that it is the right part and not damaged and that the package contains the right quantity of parts. There are other details, but these are the key details to receiving.

First, the manager only allowed deliveries between 10 a.m. and 1 p.m. You read that right. There was a three-hour window in which a delivery was accepted. If you were outside that window, you waited. Again, don't do this. This organization was very high-performing.

At exactly 10 a.m., and with military precision, everyone assigned to the storeroom on the day shift (including the manager) swung into action. Two people were assigned the duty of placing the skate conveyor in place to unload the truck, and others stood on their "marks." Literally numbered marks on the floor. Everyone was ready to perform the first step in the choreography of receiving.

After a specific truck was unloaded, the next truck rolled in. When there was a pause in trucks, or at the end of all the trucks, all the associates moved to mark #2 and performed functions associated with this new location. Mark #2 was usually the reconciliation of the bill of lading with the Purchase Order phase, or in this case, "station." Then shift to mark #3. Mark #3 was the station where the team put the delivered items in their proper location. Some items were put on the shelf as replenishment items, but most went right into job kits.

All this activity was fluid and performed through pure muscle memory. Every item was delivered, received, accounted for, inspected, and put away in less than three hours. Hundreds of items each day. I asked the storeroom manager why she was so insistent on a speedy receiving process, and her answer was very poignant: "You can't sell a part that isn't on the shelf."

It is likely that many reading this book take delivery of their storeroom items at a plant or facility receiving dock. You know far too well that items have a way of disappearing between the delivery dock and the storeroom. When items do make it to the storeroom, they may sit on a pallet waiting to be processed until the action at the service window quiets down. This is always the case for one-person storerooms.

Central to the Receiving process is the confirmation that the part received is the part that was ordered. I can't stress this point strongly enough. When a part is received and reconciled with the purchase order, the clock starts on the payment to the supplier. Failure to pay the supplier on time will result in a call that almost everyone reading this book has received: "Can you please check and see when they are going to pay my invoice?" Have you had that call? Yet?

Knowledgeable people are necessary for the reconciliation process. When storeroom orders are delivered along with the rest of the "mail" in central receiving, the errors can mount up because most receiving department folks aren't as familiar with spare parts as a storeroom associate.

Parts are received and confirmed that they are the correct parts, in the correct quantity, and that they are not damaged or do not appear to be damaged. If any questions linger on this topic, the part should be inspected by a qualified maintenance technician or reliability engineer. Don't rush into accepting a part or container that is obviously damaged. It might appear as a damaged box or even an intact box, but the part might still be suspect. Figure 2-10 illustrates this point in the activity steps to be taken as part of the Receiving workflow.

Though it might not seem like it in your practice, time is of the essence. The idea is to receive the part from the vendor, confirm its serviceability and soundness, and move it to its proper location, swiftly. Replenishment items are quickly restocked, and items destined for planned jobs are placed in the relevant job kits. It could be that the part received for a planned job is the first part to be placed in a job kit. In that case, the kit needs to be created or built. Once the final part for a job kit is received, the maintenance planner is notified that the kit is complete.

Speed is essential, safe movements of course, but as it was said, "You can't sell a part that isn't on the shelf."

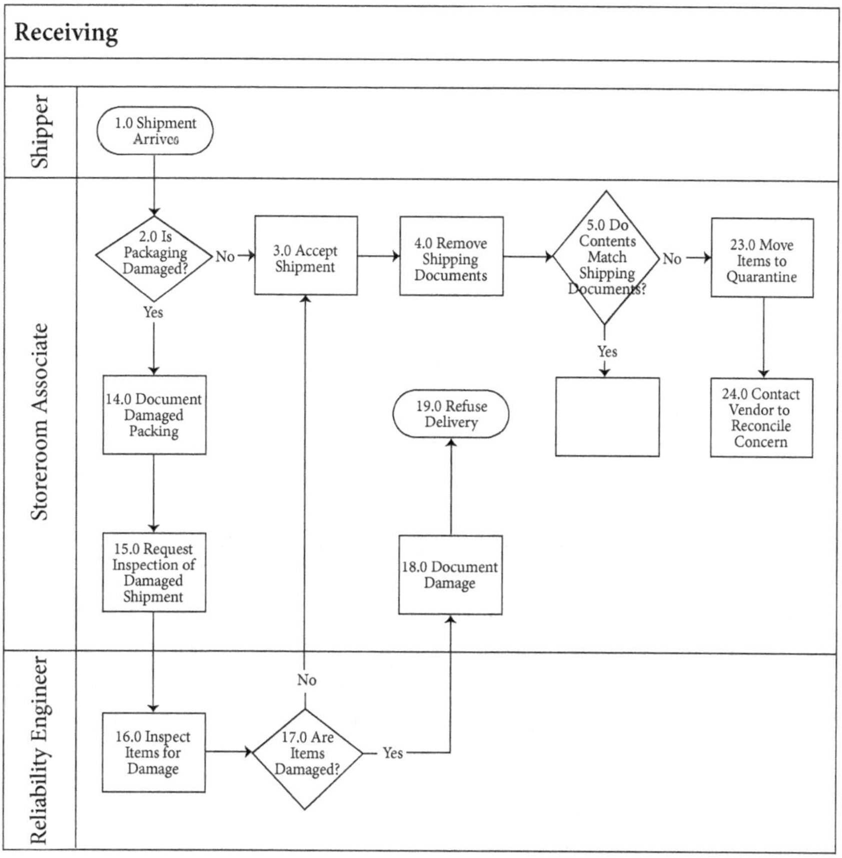

FIGURE 2-10 Confirming part serviceability.

Follow the logical path that we are laying out on our map for our journey. Parts are established in inventory; parts are ordered to fulfill that obligation. Those parts are received and reconciled, and the Foundation-Effective zone will be completed with the inclusion of the Issuing process.

ISSUING

No work order, no part. Is that simple enough? Use Figure 2-11 as a constant reminder, as this requirement is part of the Issuing process.

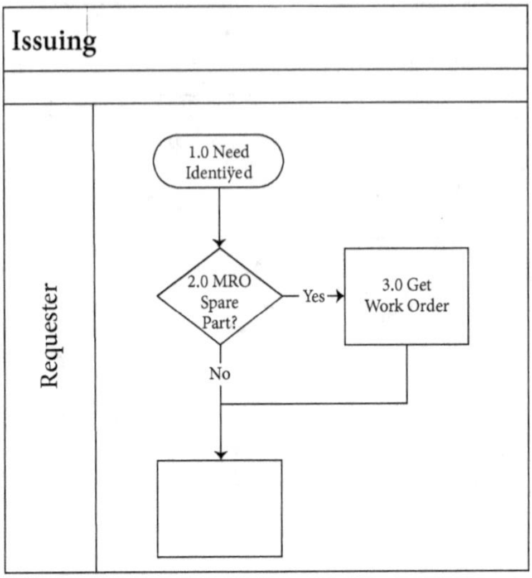

FIGURE 2-11 No work order, no part.

But what if it's an emergency? There is a good chance that you might not like my answer. If you have an urgent, or emergency, situation, you should have a subroutine in your work management system to guide you. Figure 2-12 reflects the earliest portion of the maintenance work manage-

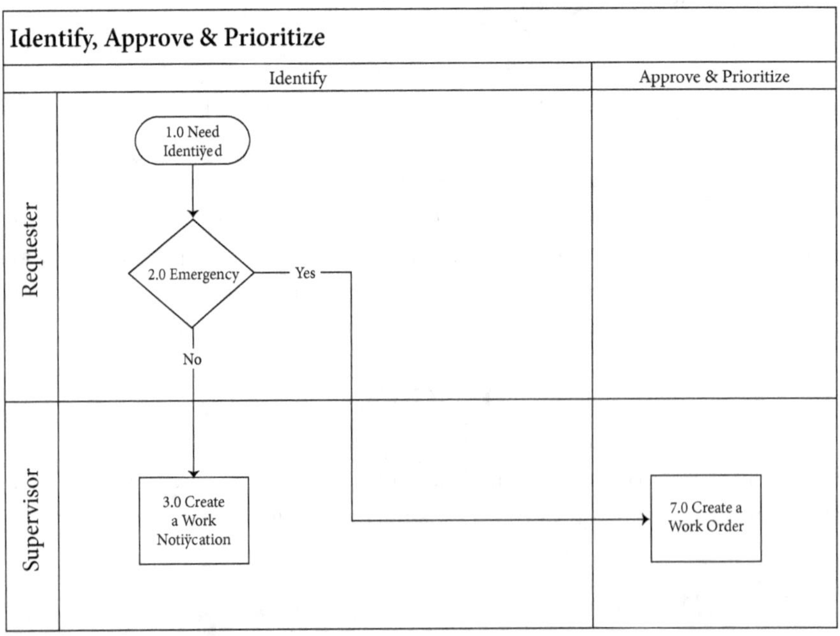

FIGURE 2-12 Identify, Approve & Prioritize process.

ment process, the Identify, Approve & Prioritize process. The IAP process was explained in great detail in my first book, *The Reliability Excellence Workbook: From Ideas to Action.* Although maintenance work management processes are not the subject of this book, it should be understood that at the very beginning of a situation that requires maintenance, the requester is asked to determine if a situation rises to the level of "emergency." If so, the supervisor is responsible for creating the work order. Otherwise, a work notification (for SAP users) or a work request is written. The work request is to be written by the supervisor.

The part I was sure would cause some eyebrows to be raised is my suggestion that if the person requesting the work is an operator or facility user, the person's supervisor is the one that generates a work order. If production calls a situation an emergency, then the production supervisor should create the work order.

Stop for a minute and give this some serious reflection. Remember, there are two metrics that require 100% adherence to be considered world class within those measures. One is that all maintenance work is accomplished on a work order. Second, you must have a work order to get a part out of the storeroom. If the work order is generated at the beginning of the process, haven't you set yourself up to comply with these two principles? As Tom Moriarty explained when defining the word "proactive," aren't you also "influencing the future"? I think so.

Parts are issued in two primary ways. One is through the service window when a person comes to the storeroom and requests a part. The other is when a kit is issued against a planned and scheduled job.

At the service window, or what you might have as an equivalent, customers must articulate their need in a way and a language that the storeroom associate can understand. It is highly recommended that customers make the request by referring to a part number. The best practice is to have the part typed on the work order. You can't get more official than having a work order with the needed parts typed on the document.

To allow the technicians or anyone else needing an item from the storeroom to communicate via part number, the technicians and others would need access to a computer and the inventory management system. If you have a sound bill of materials, you increase the odds of a technician locating the needed part on the computer (remember Rule #2?).

I want to use a metaphor to demonstrate the exchange that takes place when a part is issued from the storeroom as part of the Issuing process. At your bank, in order to take money out of your checking account, you would need to present a check made out to "cash" or insert your debit card and type your PIN number into the bank's control interface. In each instance you are providing the same confirmation that a work order would provide. This is your authorization to take money out of your account. You give the teller your check, and the teller gives you the money. There is a transaction taking place. A goods movement.

Issuing a part is almost exactly like that scenario. Issuing is the transfer of ownership of that part, with a justification for the issue (the work order vis-à-vis the check). The storeroom associate, much like the bank teller, is going to document the goods movement. The bank, the money. The storeroom, the part.

As mentioned within the element of Order Parts, from time to time any issue can trigger a reorder point. If a part is removed from the storeroom without the transfer of ownership—for example, the part is "taken"—then the trigger will not be tripped when it should be. This could result in a failure to restock an item, which could then lead to a stockout.

As you continue to think about how all this fits nicely into a very well-tuned operation, give some thought to what happens when a person asks for more items than the person needs. This puts an artificial demand on the part and could trigger a reorder when one is not necessary. Especially if the technician returns the excess parts using the Return to Stock process (discussed later).

The Issuing process is the last element in the first zone. The first zone is what I consider to be the most important. I caution clients that they should not venture out of this zone until they know what they are doing. This is the foundation. In fact, it is the Foundation-Effective zone.

SUMMARY

The Foundation-Effective zone is the *start* of it all. This is where the journey starts relative to establishing where you are. If you can't get this right, it is unlikely that you'll get anything right.

I strongly encourage all who are seeking a higher-performing storeroom or simply want a storeroom that provides service and convenience to first look at the storeroom's organizational aspects to ensure that all the pieces are in place.

First, and I'd argue the most important, is the Defined Space. For everything that is going to come after . . . when I say "the storeroom," I mean this area. You own the storeroom; you have absolute authority and jurisdiction over this space. When defining the space of the storeroom, it is practical and essential that you also establish the space for functional use. A well-running, effective storeroom should be laid out to accommodate the activities that are going to be executed in that space.

That is followed quickly with Security. Like money in the bank, you can only guarantee the accuracy of the money inventory to the extent that you can guarantee the security of the money.

Someone should oversee the storeroom. A Physical Organization must be complete with written and communicated Staffing Roles and Responsibilities. There are examples of international precedence that mandate that roles and responsibilities be well documented and shared.

There are four processes (recall what a process is from the definitions listed in the introductory chapter, "The Journey Begins") that are the backbone of this zone:

1. New Item Setup
2. Order Parts
3. Receiving
4. Issuing

The details are too numerous to list in this short summary. What is worth mentioning and considering is the order in which these four processes are listed and explained.

There is a formality to getting an item put into inventory. It should take some effort, and the requester should have some skin in the game. There is a form; fill it out. Anyone can request that an item be put into inventory. The maintenance department tells the storeroom what to stock and in what quantity. In writing. The person who carries the form to the storeroom is the planner.

Parts must be ordered or purchased in order to fill the inventory and fulfill the storeroom's obligation. Parts are ordered as part of establishing new item inventory levels, replenishing existing stock, or completing a job kit. Direct-order parts will be ordered through the work order. No more dedicated shelving for items ordered with or without a work order. This very poor practice has gone on long enough.

Receiving must be choreographed to ensure parts arrive from suppliers and are on the shelf to sell as soon as possible. Parts are only received as initial stock to replenish stocked inventory, or they go directly to a job kit. The reconciliation of the bill of lading and the purchase order starts the pay clock for the vendor.

Issuing is the transfer of ownership of a stocked item. The transactional document is the work order. No work order, no part. That should be easy enough.

WORKBOOK

I promised a consistent chapter ending. One that ends with a workbook and additional resource materials. So here is the assignment for Chapter 2. Simply fill in Table 2-4 with a Y for yes or an N for no.

TABLE 2-4 Answer Y or N

	Y or N	**N/A**
Do you have formal process guides created for the following?		
New Item Setup		
Order Parts		
Receiving		
Issuing		
Do you follow the processes as written?		
Is your storeroom space formally defined?		
Is your storeroom set up for workflow and success?		
Do you have a storeroom organizational chart?		
Is the chart up to date and accurate?		
Are there physical barriers separating your storeroom from other spaces?		
Do you have formally written and communicated roles and responsibilities?		
Are the elements in the Foundation-Effective zone working for you?		

As a regular feature of chapter endings, I mentioned that I'd be listing links to very short videos on the topics discussed in the chapters. These are videos that have been published on my YouTube Channel, *Dr. John's Maintenance Minute*. I understand through my travels that many folks cannot access YouTube on their company computer, and for that reason, I've published these links here, in print. These are links to the videos on my website; they are the same videos as on YouTube. I realize that a printed URL is not the easiest resource to use. Check out the videos in Table 2-5. I think you'll find them to be quite useful and very valuable.

TABLE 2-5 *Dr. John's Maintenance Minute* Links

Title	**Maintenance Innovators, Inc., Website Links**
Add to Stock	https://maintenanceinnovators.com/aiovg_videos/add-to-stock/
Ordering	https://maintenanceinnovators.com/aiovg_videos/ordering/
Receiving	https://maintenanceinnovators.com/aiovg_videos/receiving/
Issuing	https://maintenanceinnovators.com/aiovg_videos/issuing/
The 7 Foundational Elements of the Storeroom	https://maintenanceinnovators.com/aiovg_videos/the-7-foundational-elements-of-the-storeroom/
	YouTube Dr. John's Maintenance Minute Links
Add to Stock	https://www.youtube.com/watch?v=ECYTZIOnJOo
Ordering	https://www.youtube.com/watch?v=nndf8NQkHJw&t=8s
Receiving	https://www.youtube.com/watch?v=rRqvvR73wMs
Issuing	https://www.youtube.com/watch?v=WlxMEOrB0k8
The 7 Foundational Elements of the Storeroom	https://www.youtube.com/watch?v=k8o6xD8dqYY

CHAPTER 3

FOUNDATION-EFFICIENT

The fact is, when you look at the best teams . . . each team has all the people on it to do everything, soup to nuts.

—Jeff Sutherland

From the initial, and highly suggested, first zone to attack, the Foundation-Effective zone, your journey could take you in any direction. I strongly advise those of you who have made it this far on the trip with me to move toward the Foundation-Efficient portion of the roadmap.

I believe that the organizational element detailed in this zone—the Stores Stock Committee (SSC)—is the most important element on the entire map. I will go as far as to say that without this particular element, you will never have the best-performing storeroom that you can have, and it is unlikely that you will have sustainable success.

A bold statement? Absolutely. But I can guarantee a smooth ride along the road ahead if you will follow this advice.

Figure 3-1 is our introduction to the Foundation-Efficient zone of our Storeroom Roadmap, and Table 3-1 provides the legend for the figure.

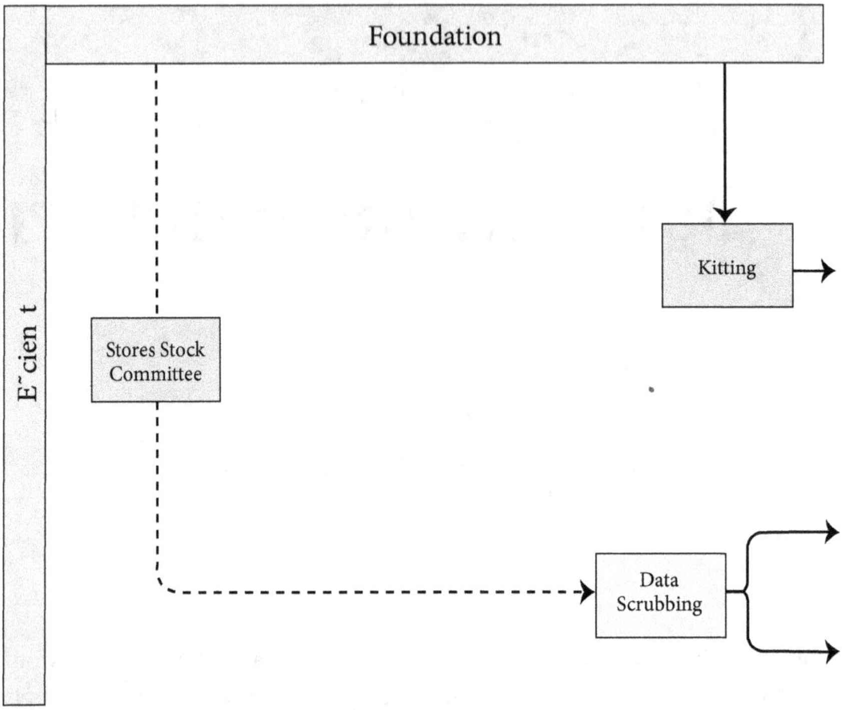

FIGURE 3-1 Foundation-Efficient zone.

TABLE 3-1 Legend for the Foundation-Efficient Zone

Element	**Activity**
Stores Stock Committee	Organizational
Data Scrubbing	Practice
Kitting	Process

STORES STOCK COMMITTEE

I'm going to go out on a limb and declare that there is not a storeroom in the world that is consistently performing at the highest level that does not have some regular engagement from the "users" of the storeroom, or those I refer to as the "victims" of the storeroom. I have made this declaration before. In *Cover Your A$$ets: Asset Management at Your Place and at Your Pace*, I wrote, "The presence of a Stores Stock Committee (SSC) is the difference maker between an average spare parts program and a world-class program. It is the secret weapon" (p. 163).

I have performed many storeroom assessments during my years as a maintenance and reliability consultant. After I present my audit findings, I host a strategy session with the client. I have never once failed to recommend forming, training, and chartering a Stores Stock Committee. This is a group of advisers who provide oversight and valuable, user-relevant input into the operations of the storeroom. The SSC doesn't run the storeroom; rather, it forms a collective of ideas and abilities to understand and help guide the storeroom.

Make no mistake. The SSC must be composed of decision-makers. This isn't a doughnut-eating, coffee-drinking group. This is the group that gets things done.

Imagine the value of having an advisory group, made up of those who can benefit from and have influence with the storeroom, engaged in ushering in the changes and improvements on behalf of the storeroom. The list of benefits and opportunities is endless.

In what is the best book for planning and scheduling, Doc Palmer writes in his *Maintenance Planning and Scheduling Handbook* a concise rationale for having a strategy to work with the storeroom. I want to "borrow" from Doc to create an equally powerful rationale for forming and engaging a Stores Stock Committee. Allow me a bit of editorial license on this. In this quote, change the word "maintenance," and substitute your company's name in its place. Doc writes, "The best *maintenance* professionals understand inventory concepts and help the storeroom optimize its inventory to achieve the best level for the plant as a whole without suboptimizing either *maintenance* or the storeroom" (p. 338).

I mentioned that I always suggest forming, training, and chartering a Stores Stock Committee. These three steps used to assemble and commission the SSC are vital, and it is incredibly important that they are accomplished in that order.

Forming a team is not as easy as it might seem. Several folks reading this book have been responsible for forming a team to accomplish a goal or mission, even if it was just picking sides for volleyball. It is not uncommon to discover that you made choices with flawed selection criteria.

The team should be made up of decision-makers, meaning the members have some authority in their own areas. Having the authority to guide people and announce (and back up) policy decisions and having a positive "bedside manner" never hurt. A suggested pool of candidates for the SSC includes:

- Storeroom manager/supervisor (the person in charge of the storeroom) as the chairperson
- Maintenance supervisor
- Production/operations supervisor
- Purchasing department/finance department member
- Maintenance planner/scheduler
- Mechanic (rotating membership)
- Electrician (rotating membership)
- Electronics technician (rotating membership)
- Safety department member (ad hoc)
- Food safety (ad hoc—if you are in the food industry)
- Environmental department member (ad hoc)

Never, ever put a manager from any other department on the SSC. I'd recommend never putting managers on any committee. The manager can kill any idea with the simple pronouncement, "We aren't going to do that." Talking about an idea killer, Amor Towles, in his novel *A Gentleman in Moscow*, describes the deftness of conversation killing best: "bringing the topic to a close with the efficiency of the guillotine" (p. 239). Reserve the managers as the people your committee are going to impress with their ideas and actions. The exception is the chair of the SSC. If your storeroom has a manager as the top person in stores, then the manager is the chairperson.

Following is a short excerpt from *The Reliability Excellence Workbook: From Ideas to Action*, about teams, groups, and committee size. But first, a quick note about committees and group sizes as they relate to clear and unambiguous communication. I have personally adopted the practice recommended in Jeff Sutherland's book *Scrum* and use it exclusively in my consulting work. He correctly observed that:

> Once the teams grew larger than eight, they took dramatically longer to get things done. Groups made up of three to seven people required about 25 percent of the effort of groups of nine to twenty to get the same amount of work done. This result recurred over hundreds and hundreds of projects. That very large groups do *less seems to be an ironclad rule of human nature. . . .*
>
> So, there's a hardwired limit to what our brain can hold at any one time. Which leads us back to Brooks [another author Sutherland researched]. When he tried to figure out why adding

> more people to a project made it take longer he discovered two reasons. The first is the time it takes to bring people up to speed. As you'd expect, bringing a new person up to speed slows down everyone else. The second reason has to do with how we think and, quite literally, with what our brains are capable of thinking. The number of communication channels increases dramatically with the number of people, and our brains can't handle it. (pp. 59–60)

The performance of the SSC is greatly enhanced when members are knowledgeable about the storeroom processes, and even more advanced if they are actually *in on it*. The Stores Stock Committee is a working committee. It is the hope for every organization that this particular group helps shape how every department will engage with the storeroom. Do you recall Rule #2 on the bill of materials and this interconnection we have with the BOM (Figure 1-6)? Doesn't it make sense that if other departments are affected (positively or negatively) by the storeroom and the bill of materials, they have a seat at the table when it comes time to create the strategy for the effective and efficient execution of the storeroom? I think it does. And should those committee members thoroughly understand how a storeroom works? Your storeroom in particular. SMRP sure thinks so, saying, "A comprehensive understanding of the entire process flow can enhance the stakeholder's ability to help control and optimize the processes" (*SMRP Guide to the Maintenance and Reliability Body of Knowledge*, p. 5).

Or as I teach in my *Best Practices in Maintenance and Reliability* course, "Stakeholders that can influence performance and safety must have a complete process understanding" (p. 141).

Forming and training a cohesive team only nets a benefit if that team stays the course of its charter—stays true to why the team formed and its mission. The Travelrat's Travel website reminds us of the importance of staying focused on the task with this old Cajun proverb: "When you're up to your ass in alligators, it's difficult to remember you came to drain the swamp."

The Stores Stock Committee team charter is a standard charter defining what the mission of the group is and the operating structure. The SSC folks are commissioned to help the storeroom by understanding the mission of the storeroom and working within their own groups to usher in change and direction.

Common goals (*Materials Management for Maintenance*, p. 395) the SSC should pursue include, but are not limited to:

- Reduce inventory quantity
- Reduce inventory value
- Improve turn ratio
- Decrease slow-moving items
- Remove obsolete items
- Stock or not to stock
- VMI/consignment

The Stores Stock Committee should meet once per month and hold its meetings to one hour. Standing agenda items would include discussions on:

- Inventory value
- Storeroom KPIs/metrics
- Items identified as obsolete and slow-moving items
- Items requested to stock or to have stock numbers created for
- Old business
- New business

This last point is crucial for sustainability. Every good thing that comes from the storeroom in terms of improved and enhanced service and convenience must be seen as being made possible because of the inclusion and engagement of the users and influencers of the storeroom as representatives on the Stores Stock Committee. There is nothing that can or will stop a team that is formed, trained, chartered, and allowed to perform at its highest level. Nothing.

As its first official act, I always recommend that the SSC define what is a critical spare part and what is an obsolete spare part. Every other part is somewhere in between those two classifications of parts. Those specific written and communicated definitions will prove helpful in the near future as you progress along the Storeroom Roadmap.

The next element is the strategic practice of Data Scrubbing.

DATA SCRUBBING

We touched on this principle before during Chapter 1, with Figure 1-9 showing two markers. In fact, the entire conversation on the Item Master Data table was relevant to this element. This is the first strategic practice we come to on our Storeroom Roadmap. Recall from the introductory chapter, "The Journey Begins," covering working definitions that "practices focus on *improvement*." "Strategic" is such an applicable word when you accept that part of a working definition for this word might be to "set yourself up for future success."

Data scrubbing is nothing more than ensuring that the information you have in your Item Master Data table for any given component is accurate, up to date, and complete. Can you honestly think of anything more *strategic* than having complete knowledge of the parts you stock? Think about the future success of the storeroom. This is very strategic.

What follows is the first test to see if you've been paying attention this far into the book. I need your absolute attention for this next portion. You are going to need to recall, with uncanny certainty, what we have covered so far.

Recall at this point that we have just formed, trained, and chartered the Stores Stock Committee, and recall who the members are, more specifically, the makeup of the committee as far as job titles are concerned. The users and influencers (aka the "victims").

Also, recall that we have as a rule (Rule #2) a complete bill of materials on all our assets. Remember that the Item Data Master table can have up to 47 data fields. All new items requested for stock will have a form filled out prior to any item being placed into inventory.

With all that prework, it is hard to believe that we would ever have holes in our documented knowledge of what is stored in the storeroom. But we do.

Figure 3-2 demonstrates the start of the Data Scrubbing practice. You should instantly recognize that the SSC commissions the data scrubbing exercise and sets the policy for how it is to be accomplished. I recommend that organizations review about 1% of their line items to determine if their Item Master Data is accurate, up to date, and complete. I'm only suggesting this be done with MRO spare parts. In this book, we are focused on spare parts.

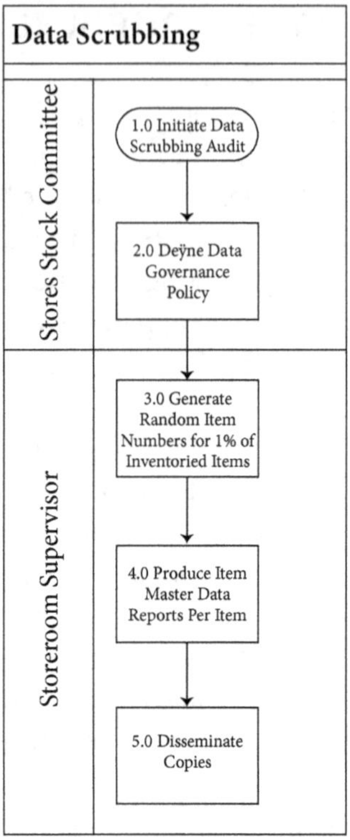

FIGURE 3-2 Data Scrubbing (partial).

The storeroom supervisor will generate a random list of item numbers and print out the data sheets on those items. The data sheets are then dispersed to the "owners of the item" for their review of the information. "Owners of the item" is a term used when referring to the agency that requested the item be put into stock. Again, if you requested it, you own it. Much like the requester for New Item Setup owns the start of that process.

Your workflow created for this element should continue to illustrate how the work flows beyond these introductory steps. A segment of the additional work to be included in your process guide might include what steps should be taken if the results from the 1% confirmation check are unsatisfactory. I suggest that further items should be reviewed if the 1% initial work returns unfavorable results. Perhaps you might specify another 1–2%.

Data scrubbing should be accomplished using the process guide flow that your team develops. At a minimum, review your Item Master Data with the Data Scrubbing practice at least twice per year until the data is routinely proved to be consistently accurate, up to date, and complete. At that time, your SSC might consider backing this requirement down to a more manageable one-time per year.

In *Spare Parts Inventory Management: A Complete Guide to Sparesology*, Phillip Slater offers this sobering thought on getting your data clean and keeping it clean: "Data cleaning and governance projects are hard to get right, and many companies have wasted a lot of money failing to do so, but clean materials master data is the foundation on which considerable business intelligence rests. Getting it right can open up a myriad of opportunities and will save you much, much more than it costs" (p. 226).

The last element in this zone is one that has confounded some of the biggest companies in the world—Kitting.

KITTING

Kitting is so easy that a 16-year-old kid at McDonald's can do it without breaking a sweat. However, some of the biggest and most successful companies on the planet have given up on the process because "it's just too hard."

"The goal of Kitting is to increase the utilization of the workforce by reducing the wasteful processes of looking for parts and waiting for parts" (*Advanced Storeroom World-Class Principles*, p. 58). This increase in utilization can be realized through the benefits of kitting jobs, which include, but are not limited to:

- Guaranteeing all parts needed for a job are available *before* the job
- Reducing/eliminating non-value-added tasks by technicians
- Allowing planners to properly prioritize and schedule jobs
- Providing more efficient use of storeroom personnel's time
- Eliminating emergency buys
- Never scheduling a job until you know that parts are on-site

The benefits and the necessity of kitting parts for planned and scheduled jobs cannot be overstated. It is imperative that organizations get this process right, all the time. Aside from planning and scheduling, the ability to prebundle work products prior to job execution is singularly the most important aspect of world-class maintenance execution. The process of Kitting is quite literally the point at which the maintenance department (top workflow) and the storeroom (bottom workflow) "touch"—as shown in Figure 3-3.

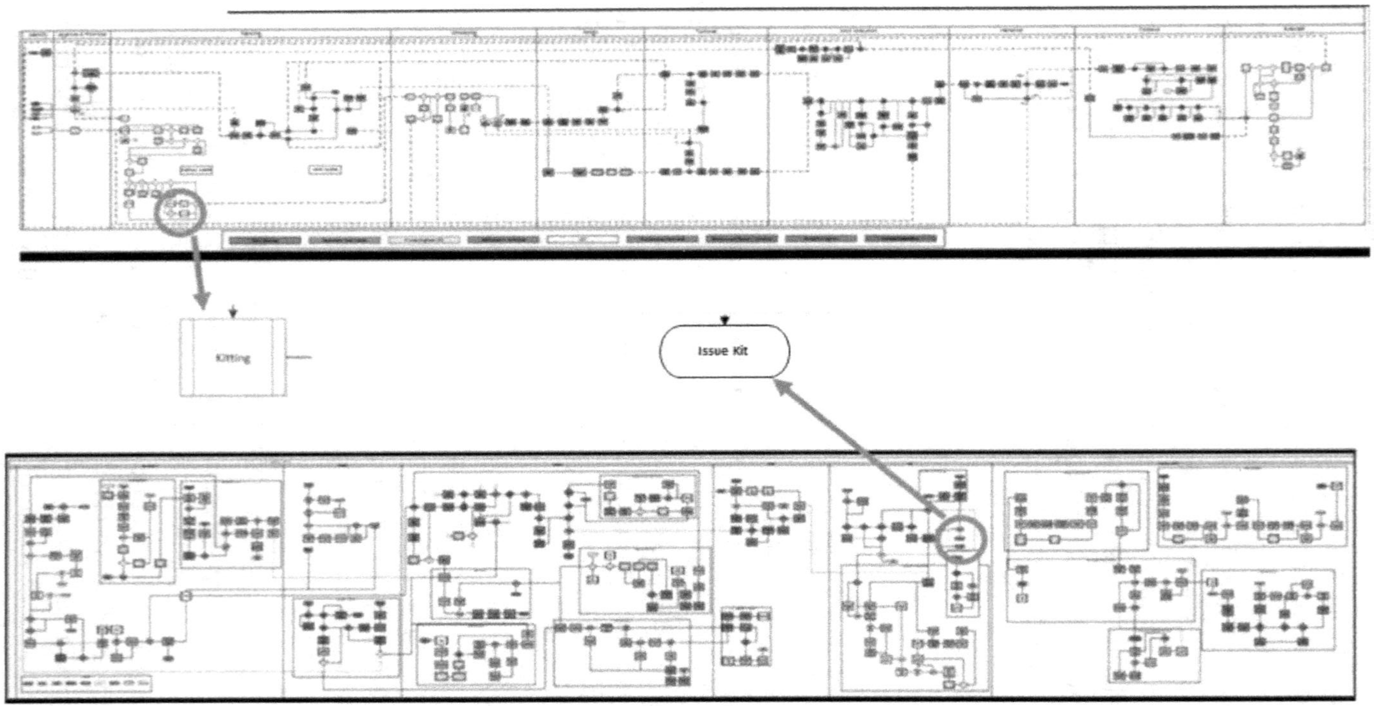

FIGURE 3-3 Where maintenance and the storeroom "touch."

Most companies that begin developing a Kitting process do so without first considering how all the pieces of the process fit together. In the simplest definition, Kitting begins when the planner identifies the parts and materials needed for a planned job. The planner communicates those items to the storeroom associate via a "pick list." A storeroom associate will bundle the parts and materials together in a kit and communicate to the planner that the job kit is complete. Simple? Not so fast.

Companies get tripped up on the following details (this is not an exhaustive list):

- How exactly is this pick list communicated?
- When does the associate pick the parts off the shelf and put them in a job kit?
- When does the inventory management system deduct the parts issue (goods movement) from the inventory?
- How does the storeroom associate communicate back to the planner when all the materials are kitted?
- Where are the kits stored once they are assembled?
- Are parts issued to the job, or is the kit issued to the job?
- Can we take a part out of the kit if the plant is down and we need that item?
- Can you return a part to the storeroom or the supplier if a part is not needed?
- How do we account for kitted parts when we are cycle counting?

There are many more tripping points, and bigger and better companies than yours have been caught up in the minutiae.

Previously, we discussed having a part number for everything in stock (and items on the BOM that we are likely to order in the future), and the benefit of having the planner order by part number instead of by part description. We've even recently had a conversation about having complete and accurate information in the Item Master Data. What we have yet to discover is the importance of knowing how your particular CMMS works when reserving items in stores for planned jobs.

I strongly encourage my clients to work with the system they have and not against it. Believe me, your life is easier if you let the machines do the heavy lifting. Determine if your computer puts a hard or a soft reservation on the part. The difference is that a hard reservation can sometimes subtract a reserved item from inventory, even though the part is still physically on the shelf. Also, a hard reservation will often put a demand on the item, and if it triggers a reorder point, the item may show up on the next day's reorder list. This is a very important detail to understand. Use the computer as the tool it is supposed to be.

Recall a previous point on keeping a low inventory and a low turn ratio. Ideally, you want to order items directly into a job kit. In the discussion on the Receiving element (Chapter 2), we spelled this out quite extensively.

If you do not have a Kitting process, start the program by kitting for preventive maintenance jobs. Start slow, and work your way up to bigger repair jobs. Crawl, walk, and then run.

Assigning a dedicated spot for job kits, once they are assembled, is a detail you want to have ironed out before you start the kitting effort. Ideally, kits are assembled inside the storeroom and then physically transferred to a location outside the storeroom. The kit storage area must be secure, too, as the parts are still technically in inventory. This is not always possible.

In Chapter 2, under the very first organizational element, Defined Space, the storeroom setup, or storeroom design, was suggested to include a "Separate job kit storage area." Figure 3-4 shows one client's very first attempt at establishing a kitting area. This is a good start.

FIGURE 3-4 Kit storage area.

Note to the reader: The chapter you are on was written weeks before this note you are now reading. While I was typing the last chapters of this book, my wife and I were having the windows in our home replaced. I'm still sitting in a construction zone as I write this note. We had 28 windows that needed to be replaced. The entire time the vendor planned and scheduled the job, we were in constant touch with the team handling the job. I was informed that the windows would arrive at the vendor so the team could confirm size, style, and operability. We scheduled a delivery and installation date. The team showed up on time, on the scheduled day, with everything they needed for the job. They are finishing up today. This is day three of the installation. This team never had to leave my house except to go home at the end of the day. They had windows, tools, equipment, associated window trim pieces, etc. This is a practical demonstration of the power of Kitting. There is also a little reference here to the power of service, convenience, and communication as well.

SUMMARY

Where the Foundation-Effective zone is the *start* of it all, the Foundation-Efficient zone is the *heart* of it all. There are only three elements in this particular zone, and every single one of them is vital to the journey.

Without a doubt, and with very little debate, the creation and success of the Stores Stock Committee will be the catalyst for sustainable growth and execution. There is not a management or leadership book written or course taught that directs an organization to limit its inclusion or to lessen people's engagement in the organization. Imagine where your storeroom would be today if you could tap into the energy of a small team whose members are committed to each other to drive improvement. If nothing else, having others understand the required workings of a professional storeroom and "lending a hand and/or ear" can be invaluable.

Make no mistake, and don't ever doubt that everyone in your organization is connected to the bill of materials. This point is so central to good order and discipline in the operation of our companies that it was discussed at the beginning of this book (Rule #2). Here in the Foundation-Efficient zone, we reintroduce the idea of a solid BOM through accurate, up to date, and complete information in the Item Master Data. This data exists in the tables within your inventory management system for everything that you have in inventory. From time to time, you must confirm that the information you have on the parts and materials meets the level of excellence, which, in this case, is information that is accurate, up to date, and complete. This practice is Data Scrubbing. Data Scrubbing is the first strategic practice we encountered on our roadmap.

There is a dragon to slay. A monster is out there that we will need to face at some point. We encounter that being here in the Foundation-Efficient zone. That behemoth is the Kitting process. Bigger companies than yours have fought the battle and are still lying in the wasteland, so traumatized that they refuse to try it again. "Yeah, we tried that once, and it didn't work." Meanwhile, 16-year-old kids at McDonald's do it every day as their regular duty. Drive up to a McDonald's and ask for a #1 value meal. I know exactly what you are going to be served.

Aside from planning and scheduling, Kitting is the most significant utilization-increasing process you will ever execute. Get this process, understand this process, figure out your CMMS, and get very good at this. It has the potential to multiply your workforce by 10–20%. It is that valuable.

WORKBOOK

Here is the assignment for Chapter 3. Simply fill in Table 3-2 with a Y for yes or an N for no.

TABLE 3-2 Answer Y or N

	Y or N	N/A
Do you have formal process guides created for the following?		
Kitting		
Data Scrubbing		
Does the planner send a pick list to the storeroom?		
Is someone in the storeroom assigned the role to kit jobs?		
Is your Item Master Data table complete and accurate?		
Do you have a Stores Stock Committee?		
If so, is it formally chartered and trained?		

Check out the *Dr. John's Maintenance Minute* links in Table 3-3. Review them with your team and use them as a valuable teaching tool.

TABLE 3-3 *Dr. John's Maintenance Minute* Links

Title	Maintenance Innovators, Inc., Website Links
Stores Stock Committee	https://maintenanceinnovators.com/aiovg_videos/stores-stock-committee/
Data Scrubbing	https://maintenanceinnovators.com/aiovg_videos/data-scrubbing/
Kitting and Delivery of Kitted Items	https://maintenanceinnovators.com/aiovg_videos/kitting-and-delivery-of-kitted-items/
Where Maintenance and the Storeroom Touch	https://maintenanceinnovators.com/aiovg_videos/where-maintenance-and-the-storeroom-touch/
	YouTube Dr. John's Maintenance Minute Links
Stores Stock Committee	https://maintenanceinnovators.com/aiovg_videos/stores-stock-committee/
Data Scrubbing	https://www.youtube.com/watch?v=NHitnBpeGi0&t=23s
Kitting and Delivery of Kitted Items	https://www.youtube.com/watch?v=yAkSndUfwVg&t=46s
Where Maintenance and the Storeroom Touch	https://www.youtube.com/watch?v=0j4J_EiFYT0&t=38s

CHAPTER 4

INTERMEDIATE-EFFECTIVE

The main function of the storeroom is to provide parts accurately when needed. It plays much more than a pivotal role; it is the *hub* (center) of successful reliability operations.

—Dr. Nathan C. Wright

This is a roadmap, and like any map, you can choose your destination and route based on personal preference or your final destination. In my family, I can find no fewer than 20 witnesses who will tell you that when you take a trip with me, you are going to do a lot of driving. I like to maximize the things that I see and the experiences that I have. That's not always the case in business. We should be efficient, but we should be effective first. Therefore, I suggest that your organization return to the effective area of the Storeroom Roadmap and begin to steer toward some processes to shore up your effectiveness.

Always check your "review mirror" and keep in mind that you've laid a lot of groundwork for future success by tackling the organizational elements, tactical process elements, and the one strategic practice element covered in the first two zones of this map. We have a full tank of gas, a clear mind, and an open map, and we are ready to drive toward what I consider to be the Great Plains and foothills, which are the Intermediate-Effective zone elements.

Figure 4-1 lays out the course ahead for the Intermediate-Effective zone of our Storeroom Roadmap, and Table 4-1 is the corresponding legend.

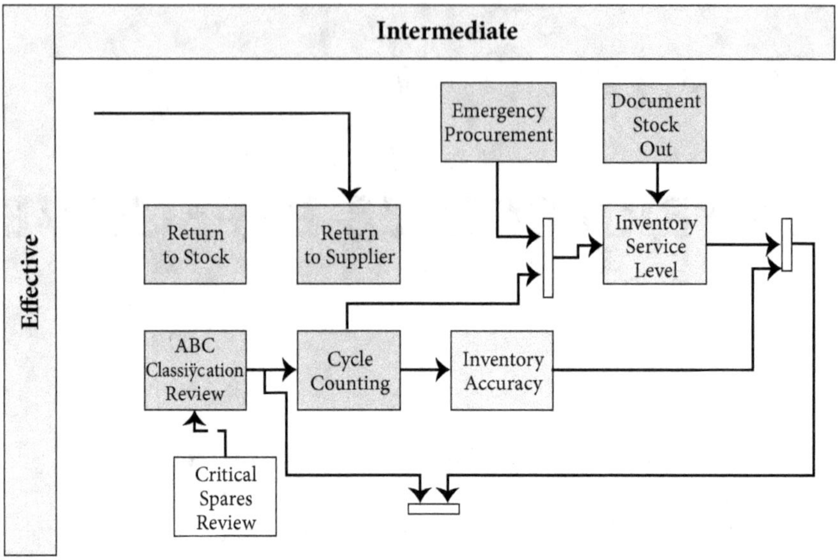

FIGURE 4-1 Intermediate-Effective zone.

TABLE 4-1 Legend for the Intermediate-Effective Zone

Element	**Activity**
Return to Stock	Process
Return to Supplier	Process
Emergency Procurement	Process
ABC Classification Review	Process
Critical Spares Review	Practice
Cycle Counting	Process
Inventory Accuracy	KPI
Inventory Service Level	KPI
Document Stockouts	Process

RETURN TO STOCK

If there were ever a process we hoped would never be used, the Return to Stock process might be one of them. This process is most often utilized when a spare part or material item is determined to be in excess to a job or just not needed. The result is that the item must and should be returned to the storeroom.

The roadmap shows this as a possible route coming from the Issuing process. Returning an item to inventory could just as easily be associated with the Kitting process. Regardless, there are several reasons for an item to be returned to stock, and there are as many problems that can occur when items are returned to stock. Again, this is not a process you want to use, but you must have this well documented and it must be a well-followed process because it will be used despite your best efforts. Probably used often.

Let's consider why an item might be returned to the storeroom to be placed back into inventory. The primary reason is that the part was not needed for the assigned job. The assigned job might not necessarily be a planned job. So our starting point always begins with either a planned (and scheduled) job, for which we hope we have kitted the necessary items in advance, or a job that wasn't planned (or scheduled). It's likely a technician received the part(s) via the service window at the storeroom.

A lack of troubleshooting skills has become prevalent in today's maintenance talent shortcomings. This lack of troubleshooting skills can lead to a technician requesting several parts from the storeroom, in hopes that one of those parts one (or more) will fix the issue. Picture a technician with limited troubleshooting skills, or limited time for troubleshooting, going to the storeroom and requesting five parts. It may be, and we hope, that one of those parts, or a combination of those parts, fixes the issue. It is unlikely, just by probability, that the defect required the removal and replacement of five individual parts. Not impossible, but how likely is that? Honestly?

If the issue is repaired after replacing three parts, we hope the technician returns the two other parts to the storeroom.

It may be that the maintenance planners haven't honed their planning skills and are facing a maintenance assignment that might need a combination of parts. They (the planners) are not certain which parts are truly needed, but due to reduced troubleshooting effectiveness, the planner includes five components on the pick list. The technician replaced two parts from the kit, and the issue was resolved. Three items are available to be returned to the storeroom.

Another possibility is that the technician decides not to replace a part that was in the job plan to be replaced and has residual stock items to return as a result. For this particular instance, a one-on-one counseling session is highly recommended. Technicians should never decide *not* to replace a part that was planned and scheduled without confirming with their supervisor.

Of the reasons you want excess items returned to the storeroom, the most important is the need to "credit" the work order. If the rules we established at the start of this book are adopted and followed, and with the mandates set forth in the Issuing process, there is a 100% probability that the parts removed from inventory for the job were issued against a work order. In the final accounting of a maintenance job, you will want to provide credits back to the work order, so the final work order reflects the actual charges to the job and a full accounting of the parts that were required to make the repair. This gives the maintenance planner a real-life costing of the job and sharpens their job plans for the next issue that arises for which a plan is needed. This is tremendous job feedback.

Another valuable reason for returning items to the storeroom is for analysis of what took place. If an organization has discipline in this process, there are many time-saving and money-saving assessments that can be made from the data collected through goods movements. This is helpful not only for asset management but also for maintenance budgeting.

Accounting for the true cost of maintenance aids in calculating life-cycle cost, and helps organizations understand when it is time to get rid of a piece of equipment that is costing too much to own.

Oh, but there is trouble ahead on our journey if we don't get this right.

Let me start this last discussion with a simple observation: You never want technicians putting parts back into the storeroom themselves. Don't do it. First, there is no reason for the technician to be in the storeroom. Second, they don't even let regular people put books back on the shelf in a library. We are taught to put them on mobile racks and the librarian will reshelve the books. After all, you can't figure out the Dewey Decimal System on your own, and you'll goof up the order of the books. Imagine trying to cycle count the inventory after a herd of mechanics has put parts back on the shelves.

There is a myriad of concerns when a part is brought back to the storeroom to be placed back into inventory. Here is a short list. By no means is this list exhaustive:

- Was the part issued from the storeroom on a work order?
- Is it even a stocked item? (Ridiculous, right)?
- Is it an item that isn't stocked, but has a stock number? (Remember the 0 min/0 max?)

- Is the item complete, with no missing parts?
- Is the item in the original packaging?
- Has a replacement component already been reordered for replenishment?

I didn't even touch on the common practice of the maintenance department "discovering" items during its 5S event in the shop and returning those old, rusted, compromised items to the storeroom. Only receive those back into the storeroom room if they satisfy the Return to Stock process.

Figure 4-2 demonstrates some activities that would be a very good start to your Return to Stock workflow.

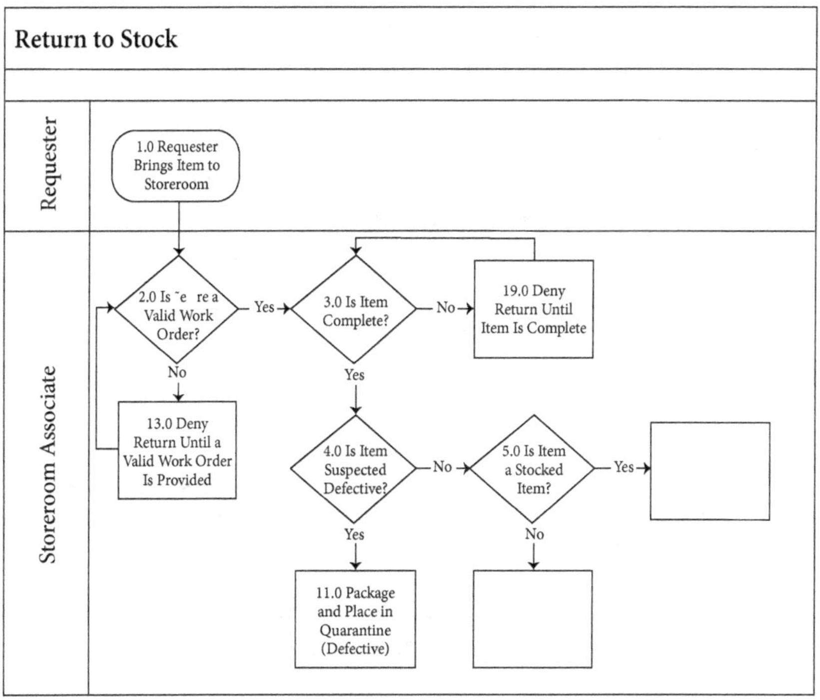

FIGURE 4-2 Return to Stock workflow (a start).

A final thought on this process. As mentioned, you understandably want to restrict the occurrences when items are brought back. But when there is a legitimate reason to request a returned item be placed back into inventory, the process should be incredibly easy for the technician. (An operator might also bring items back to stores.) The more complicated the process, the more scrutiny the employees receive, the less likely they are to return the material. They will likely throw it away, resulting in an added cost to your company. Or worse, keep it in their toolbox for use later. That could be a very dangerous decision.

RETURN TO SUPPLIER

Compared with Return to Stock, the Return to Supplier process is an element you are likely to use quite often, and one that is usually beneficial to your organization.

Similar to the previous element, there are several reasons this particular process might be called into play. As it appears on the roadmap in conjunction with the Receiving process, it likely becomes necessary as the result of having an item received that was the wrong item, was damaged, or was undesired for some other reason. The Return to Supplier process is always the first step to take when dealing with inventory identified as no longer needed because of an obsolescence review and discovering that some good and valuable parts are no longer needed to support the company's reliability efforts. These parts should be returned to the supplier for a possible cash refund or other financial benefit.

Returning items to the supplier always requires prior approval from the supplier. This is accomplished by asking for and receiving an RMA (return merchandise authorization). The RMA alerts and prepares the supplier to be ready to receive the part.

There are some significant rules regarding returning parts and materials to the supplier. Adhering to all of these will keep an organization in good standing with its vendors:

- Ensure the part is complete, safe, and serviceable.
- Make sure the part is in the original box or container.
- Make sure the part is clean.
- Make sure the part is new and has not been used.

The purchasing department can sow a lot of goodwill by addressing the potential for returning items to the supplier when establishing suppliers as *approved* vendors. A short conversation and documentation on a memorandum of understanding can pave the way for a smooth return process. Preferably, an agreement that requires no "restocking" fee.

Figure 4-3 outlines some key features of the Return to Supplier process.

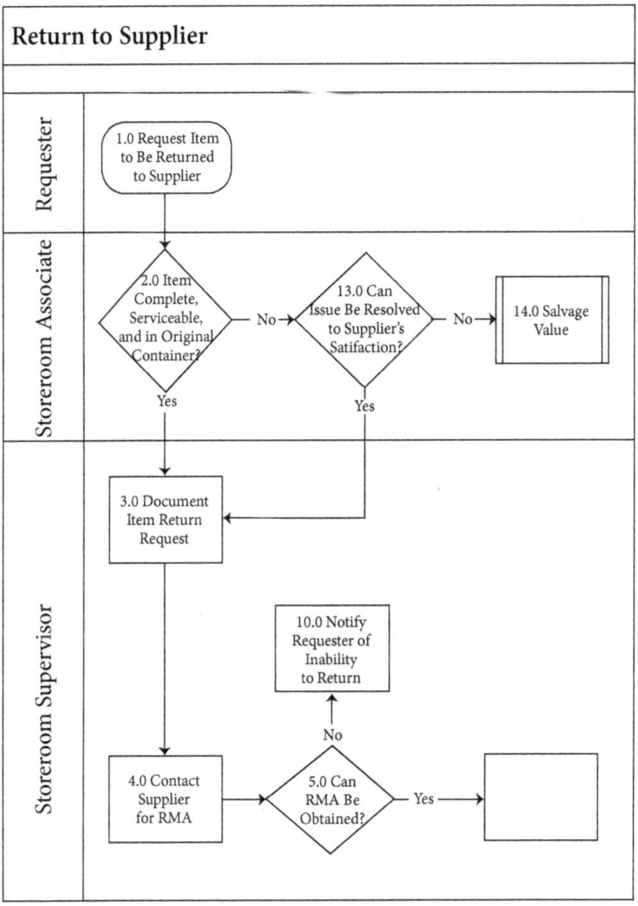

FIGURE 4-3 Return to Supplier process (partial).

EMERGENCY PROCUREMENT

We have developed a bit of a theme in the early part of the Intermediate-Effective zone by mating elements in this zone with their typical "opposite" elements in the Foundation-Effective zone:

- Return to Stock—Issuing
- Return to Supplier—Receiving

The Emergency Procurement element typifies the kind of behavior you'd like to eliminate if possible. There will always be a need to purchase items in urgent or emergency-type situations, but you must track these events and understand what led to a crisis decision in the first place. Anecdotal evidence points to a best-in-class target value of <1% for purchases being made as emergency buys.

The last process mating is:

- Order Parts—Emergency Procurement

As with most situations, people don't make their best decisions in a crisis; they make their most creative decisions. Top-performing organizations want creative thinking for marketing, research, and manufacturing. When it comes to asset reliability, it's better to maintain an even keel and perform maintenance by the book. It is, however, impractical to believe there will never be a need to rush a part in. But at what cost? Sometimes the cost of a shipment exceeds the value of the part itself.

You need to have a contingency plan for this situation. That plan must require the proper vetting of the need. A technician or requester can establish the urgent need for a part, but the supervisor must vet it as a bona fide emergency. Failure to perform this crucial "double check" can cause maintenance spending to spin out of control and lose credibility with the storeroom and purchasing.

To put this into perspective, the Emergency Procurement process begins when the requester needs an item, but the required item is not available for routine issue in the quantity needed at the time needed. The *need* is elevated to the requester's supervisor, who must determine if the item is truly needed for an issue that compromises personnel safety, asset safety, production, or the environment.

Figure 4-4 is used to illustrate the vetting of an emergency situation.

There are a number of hidden sins within this seemingly innocent element. Chief among them is to turn a blind eye to the occurrence and write it off as the cost of doing business. Or worse, have a "We've always done it that way" mentality.

Think back to the last time you needed something at home or ran out of something you needed. Was it a benign event? What was it? Bread? Milk? Toilet paper?

I said before that believing there would never be an emergency procurement need was impractical. I stand by that and have defended the practice for decades. There is just no accounting for everything you could possibly need and having it on hand when you need it. We've all run to the store to get that one last item. Everybody.

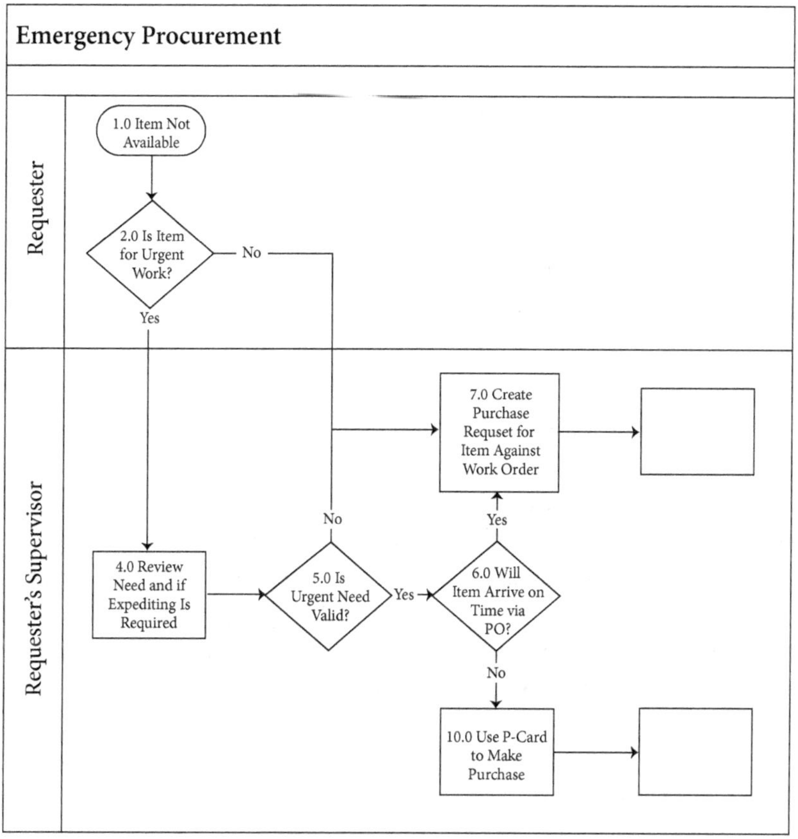

FIGURE 4-4 Emergency Procurement process (partial).

Here is a real-life scenario and one that you are probably facing today.

A moderately sized manufacturing plant may have 1,500 to 2,000 identified and documented assets. Let's go with 1,500. Of those 1,500 assets, each one is likely to have about 10 individual components on it that make it work and perform its function. For example, a conveyor might have a motor, coupling, gearbox, bearing, rollers, belting, sprockets, chain, controls, etc.

Imagine 1,500 assets and 10 components each, which nets you 15,000 uniquely important components.

It is most common for a facility of this size to have an inventory line-item count of about 1,200 and an inventory value of around $1.4 million.

The same facility will undoubtedly have 400 unique PM work orders.

To summarize quickly:

- 1,500 assets
- 15,000 uniquely important components
- $1.4 million in inventory over 1,200 parts
- 400 unique PM work orders

There is a high probability that an important part will break and you won't have it. The math works against us. We need a backup plan. An Emergency Procurement plan.

Hear me out though. It ought to hurt, just a little, to activate the emergency plan. After the dust settles, someone is required to answer for what happened. Smart people need to figure out how to keep the situation from happening again.

Figure 4-5 first appeared in *The Reliability Excellence Workbook: From Ideas to Action* (Ross, p. 274) and was used to illustrate the approximate placement of the maintenance work management activities superimposed over the classic P-F curve. This is not a book on preventive maintenance or a book dedicated to general maintenance and reliability, but this figure is useful to demonstrate an important point.

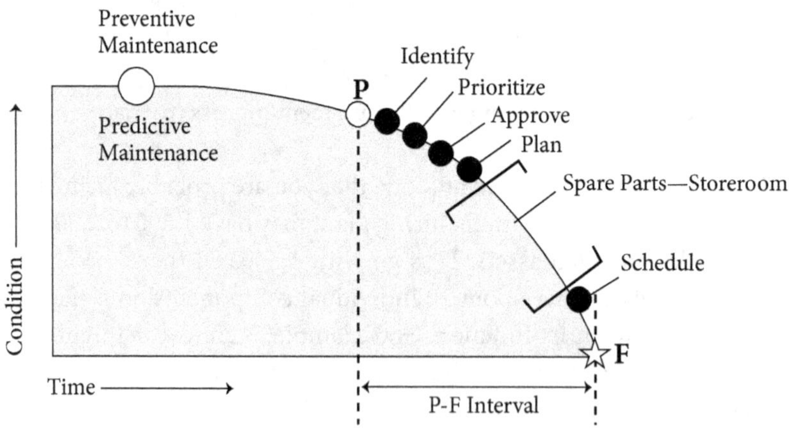

FIGURE 4-5 Maintenance work management and the P-F curve.

While contemplating the image shown in Figure 4-5, consider the possible answers to these questions that might be asked after the dust has settled around an emergency part purchase:

- Was there PM or a PdM (predictive maintenance) executed for this component?
- If not, why not?
- If there is a program, has the PM or PdM been shown to be ineffective?
- On the P-F, around the point marked "P" for point of potential failure, was this an event that a trained and engaged operator could have discovered?
- If not, why not?
- Should it be?
- If the job was planned, in the Spare Parts portion immediately after the Plan node, was the part identified and ordered?
- If not, why not?
- Should it have been?
- Was the part that was rushed in a part that is stocked?
- If so, why was it not in stock?
- Should we have been able to detect a defect in time to order the part normally with a PM, PdM, or AM (autonomous maintenance)?
- If so, why didn't we?

I suggest a short meeting after the fact. Let the dust settle around whatever the emergency situation was. There should be an accounting of what happened and how it can be avoided. Take an opportunity to turn lemons into lemonade. This is not a punitive meeting but rather a meeting to determine if there were other options and how another such occasion can be avoided. The propensity for knee-jerk reactions is very strong in this category. How many bad situations are improved by ignoring them and not addressing them?

If we encounter a shortage in the heat of battle, and it is further perceived to be dire, we are inclined to make rash yet permanent decisions. For example, "We need one, better order two just in case." (See New Item Setup in Chapter 2.) Worse, technicians are likely to keep an extra part in their personal stash.

Emergency Procurement. You don't want to venture into this element on your trip, but if you need it, you better have it laid out and ready for use. Why? Because you're going to need it.

ABC CLASSIFICATION REVIEW

If an element on the roadmap could ever be underestimated, ill-used, misunderstood, and just flat-out ignored, this is the one. Clients who use the ABC classification usually speak to their practice with an air of confidence. It is just as likely that an organization doesn't use it at all. That second group is missing the value of this great process. The first group, if I were ever to ask them what the process meant, or what they did with the classifications, I'd surely be met with a blank stare. In short, most organizations are getting the same results—not much.

So despairing is ABC analysis that Phillip Slater marked it as the fifth of five common supply chain techniques you *should not* use when managing spares. Slater writes of all five techniques, including ABC analysis:

> Putting aside all the technical reasons why these techniques don't work, there is one very good reason why you should not try to apply these techniques: employee confidence. Attempts to apply these techniques will cost you time, effort, and money, and because they don't work effectively with spare parts, they will also cost you the trust that your spare parts management system can deliver the parts required when needed. And that might just be the greatest cost of all. (p. 15)

As I wrote just a few sentences before, many organizations don't use ABC classification and therefore don't do any ABC analysis. Don't force this. This is a powerful process, but not one that should be approached half-heartedly.

The ABC Classification Review, better yet, the successful execution of the principles of ABC classification, has the ability to set you and your team up for some *open field running* on your journey. I suggest that you journey through this element. To do so, you must understand and use it to your company's advantage.

There is a companion piece to this process, and it resides in the Intermediate-Efficient zone. This other process is the Adjust ABC Classification Model. This particular element will be discussed when we get to that zone.

ABC classification was first introduced in this book in Chapter 1 while explaining Rule #3—Complete Item Master Data. The tables that make up the Item Master Data in your inventory management system typically have a field to record the classification of the part. This may not be a feature on lesser or homegrown maintenance management software. More advanced CMMSs and their internal inventory management systems will likely have a preprogrammed algorithm that calculates ABC based on cost, frequency of issuance, and perhaps asset criticality. This is an algorithm that is often "turned off" or just ignored. Check the capabilities of your system. If possible, you should try to have your inventory management system perform the calculation for you.

Maybe we should start by discussing what the ABC classification is. Would you agree with me that the assets in your facility differ in their level of criticality? That is, some assets are more critical or important than others. I hope that makes sense, because that underlying truth guides just about everything we do in maintenance.

If you agree that the assets within your operation have a competing level of criticality, doesn't it further make sense that the components in the storeroom that are stocked in support of the assets in the field would then also have some stratification in terms of their importance?

The ABC Classification Review element reviews the classification process to determine if you have the correct "spread" and if a component has changed classifications based on the criteria for which classifications are determined.

You've already accomplished a lot of work in preparation for this element in the intermediate levels of the roadmap. You did the work back in the earlier portions of the map. The BOMs that you put together (or are working on) and the insistence that items in stock are in support of an asset and that everything is "on the system" position you perfectly to understand and execute this process.

The ABC Classification Review element in this intermediate zone is your opportunity to identify components in the storeroom and rank them in comparison with each other and against an accepted criteria. This information will continue to provide context in every storeroom decision you make going forward from this point on the map.

The good news? There are infinite ways to establish a classification system for spare parts. The storeroom associates are not required to make this decision on their own. Your storeroom now has the collective wit and wisdom of the Stores Stock Committee. Remember, the SSC is made up of decision-makers. In this process, the committee and the entire facility can determine how *important* a component is, relative to other components in stock.

First, let's work to understand ABC classification.

In his preeminent work, *Maintenance and Reliability Best Practices*, Ramesh Gulati gives us some helpful instruction, starting with this preamble: "The inventory items are divided into three categories (A, B, and C), according to a criterion established by the organization" (pp. 145–146). He further discusses the analysis of the inventory and uses the phrase "inventory stratification" (p. 124). "Stratification" is the perfect visual to have in mind. There certainly are layers of spare parts, and the boundaries among those layers tell us how much we care about the part.

At its most basic, an A item is more important than a B item, which in turn is more important than a C item. The SSC will determine what characteristics will be used to calculate and determine if an item is A, B, or C. There are any number of characteristics that could be used, such as:

- Cost
- Frequency of issuance
- Safety impact
- Environmental impact
- Component lead time
- Asset criticality

The fact that any manner of characteristic can be used to determine the classification adds to the confusion of how to classify a part. Since this is a *review* process, the criteria for dividing parts have previously been determined, and you are simply applying the criteria to the storeroom movements and operation.

The Stores Stock Committee commissions the review of the Item Master Data, specifically the characterization of the parts (A, B, C) to determine two outputs:

- Should a part be moved in classification based on the criteria (and possibly physically moved in the storeroom)?
- Is the spread among A, B, and C sufficient based on criteria?

Here is where it can get confusing.

Of all the characteristics that can be used to determine an A, B, or C classification, Gulati settles on a combination of the item's value and usage. Does that sound familiar? It should. It is likely the algorithm your CMMS for inventory has in place. The popularity of Gulati's book and the general manner in which the maintenance and reliability industry accepts its precepts can be a bit perplexing as different weights can be assigned to the value of the part vs. the frequency of issue of the part. This often requires a healthy debate within the SSC.

Table 4-2 summarizes Gulati's research on what is most commonly found in terms of the spread in the rankings of components (p. 124). Curiously, he doesn't tell us if these are world-class or even correct values, just those that are commonly seen. His inclusion of these values in his book might imply that he is endorsing them as a best practice. Regardless, these values and ranges could generally be considered ideal for most operations, as attaining these values would favorably position a company to use the classification as a management tool.

TABLE 4-2 ABC Distribution Spread

Category	% of Total Inventory Value	% of Total Inventory Volume	Average % of Volume
A items	60–80%	10–20%	15%
B Items	15–25%	20–35%	25%
C Items	5–15%	55–75%	65%

Figure 4-6 is a reproduction of a graphic representation of Gulati's work (p.124).

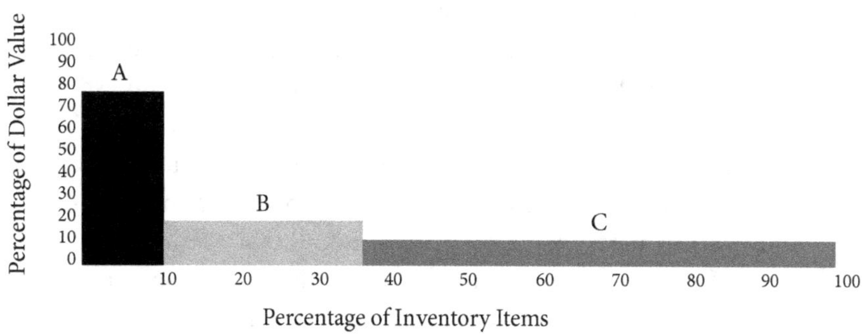

FIGURE 4-6 ABC analysis.

Some companies have found success in establishing the ABC classification with only the frequency of issue as the determining factor. A items would tend to be those that issue the most frequently, B items next, and so forth. This might lead to tighter stratification, since it would understandably be easier to achieve and monitor.

In its *MSC Materials Management for Maintenance* course materials, the Marshall Institute mentions that A items should represent about 15% of the inventory, B items about 35%, and C the remaining 50% (p. 518).

There is a small division in the ranks of the storeroom purist who wrestles with defining ABC beyond cost and the number of times the part issues out. A further divide of the maintenance and reliability community comes in the juxtaposition of Gulati's work describing A items as "one-of-a-kind parts with long delivery, high cost, and low demand" (p. 124). Possibly akin to a critical spare part. Other sources provide guidance that A items are routinely or highly used items with a high frequency of issues. Can a part be both? I'm in the camp that suggests that a part that is one of a kind and very critical is a part that you don't want to be issuing out, ever.

Gulati provides a very real and very good example of his understanding of parts classification (p. 125). Remember, this is just an example, and Gulati is not suggesting that these are the boundaries:

- A items = value over $1,000 and usage rate less than 6/year
- B items = value of $100–999 and usage rate over 6/year
- C items = value below $100 and usage rate over 12/year

Again, this is just an example, but it shows how "usage" coupled with "costs" might be broken out and stratified over several parts classifications.

The goal should be to use an algorithm that accounts for the frequency of issue and cost of the item for the calculation of A, B, or C. This accounting scheme will balance the purchasing department's main interest, which is likely the cost, with the maintenance department's main interest, which is having the part "there" when needed.

Figure 4-7 represents the ideal start to the ABC Classification Review.

At some point, we should discuss why it's important to have ABC classifications for our parts in the storeroom. Recall that at the beginning of this section, I opined, "If an element on the roadmap could ever be underestimated, ill-used, misunderstood, and just flat-out ignored, this is the one." ABC classification is underused and misunderstood mainly because no one knows what to do with the information.

You want to stratify your inventory into some agreeable categories for many important reasons. Likewise, there are many important reasons to secure a proper spread among the different classifications of parts.

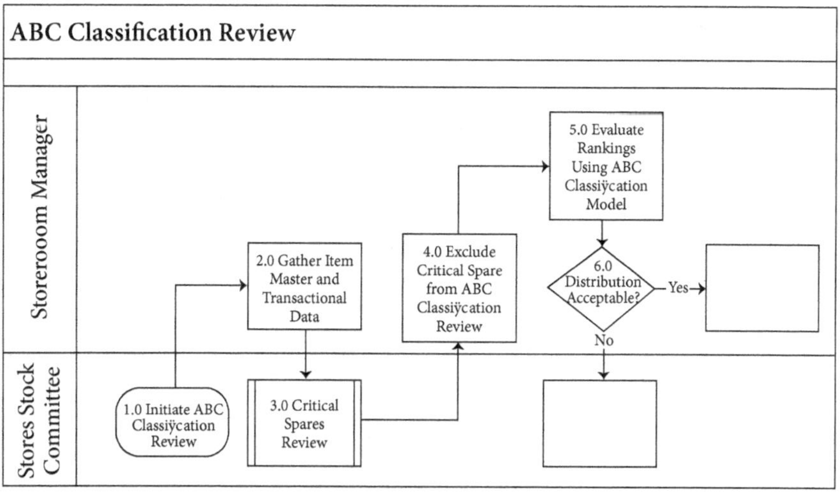

FIGURE 4-7 ABC Classification Review (partial).

From the financial perspective, dividing your spare parts into the A, B, and C categories can help you to manage and even reduce your inventory value.

If A items are the lowest quantity of items, yet hold the greatest collective value, that would be important information to know. For example, if your current inventory value was \$2.4 million, and you had a mandate to reduce your inventory by \$500,000, you would have some very tough decisions to make. Using Ramesh Gulati's numbers shown in Table 4-2, between 60% and 80% of the value of our inventory could be held within A items alone. Using a middle-of-the-road value of 70% would make A items worth \$1.6 million, leaving only \$720,000 for B and C items combined. If this hasn't hit you yet, you would need to eliminate almost 70% of all B and C items to account for the \$500,000 to meet your reduction goal.

It should be clear to you that one of the benefits of tracking A, B, and C is to aid with managing inventory value. As Gulati states, "The objective of this analysis is to move items from category A to B and from B to C in order to minimize inventory costs" (p. 125).

Another lesser-known value to having this stratification is that it helps with the reliability of your facility. If A items are important, and they are, shouldn't those same parts, the actual ones in use out in the production effort, have some preventive or predictive strategy attached to them? The strawman argument against such action would be admitting that you have a part that is so important in the storeroom to be classified as an A item, but the actual unit attached to a production asset has no real value. Does that make sense?

What components in your storeroom are more likely to be insured or to be cycle counted more often? A's, B's, or C's?

If you only have one person in your storeroom, should that person spend as much time caring for a tube of caulk as for a 50-hp motor?

Understanding the ABC classification and what to do with the stratification of parts can propel the business savviness of your storeroom to another level. Having an engaged team of stakeholders (the SSC) to commission and facilitate a regular review of the storeroom's program will provide a greater understanding of what the storeroom stocks and in what quantities the storeroom carries that stock.

CRITICAL SPARES REVIEW

Do you know that old saying that starts, "If I had a $1 for every time . . . ?" That's how I feel about the little practice of Critical Spares Review tucked into the Intermediate-Effective zone, the dotted line attached to the previously discussed element, ABC Classification Review. The misunderstanding of critical spare parts is so pervasive in the global maintenance and reliability community that it is almost impossible to determine what a critical spare part is. Almost.

Take just a minute to look back at Figure 4-7. In that figure, you'll find that item 3.0 in the workflow is a box. That box represents a singular activity, as do all boxes in workflow symbology. That step indicates that the Critical Spares Review practice should be called into play during the ABC Classification Review. A quick review: From the symbology in Figure 1-1, we know a rectangle with parallel lines on the sides indicates another defined process. We also know from the example process guide provided in Chapter 1 that a touchpoint in a workflow is the point where one process touches another. The Critical Spares Review is a touchpoint for the ABC Classification Review.

As we have defined a practice, or more specifically a *strategic* practice, as an activity that is likely to be executed no more than four times per year, it is up to the organization to determine how often it would like to run its potential critical spare candidates through the labyrinth of criteria to confirm a part's criticality. "Labyrinth" is a strong word. The process for calculating the criticality of a part really is not tortuous as the term would make you think, but it does make you *think*.

There is a companion piece to this practice, and it resides in the Intermediate-Efficient zone. This other process is the Critical Spare Parts Algorithm. That element will be discussed when we get to that zone on the map.

This is another review that the Stores Stock Committee commissions and provides guidance for and the resources to complete. If you are keeping score, this is the third time the SSC has put the ball in play by commissioning a large-scale review: Data Scrubbing, ABC Classification Review, and now the Critical Spares Review. The SSC is a very important group. You can see why truly successful storerooms are partnered with a stakeholder group that is engaged in the functional overview of the storeroom's duties and requirements.

I want to dispel a belief that most people mistakenly have. Some people believe that to ensure we always have a particular spare part in the storeroom, we must call it a "critical spare part." They further believe that a part that is "critical" must always be in the storeroom. None of that is correct, nor has it ever been correct.

A part is determined to be critical if it rises to the level of criticality, based on enumerable characteristics. Those will be discussed in the Critical Spare Parts Algorithm element.

At the execution of this practice, here in this zone, under the interpretation of the ABC Classification Review, the SSC is chartered to set aside spare parts that might be successful candidates for inclusion in the very exclusive club that is *critical spare parts*. It should be mentioned that true critical spare parts don't have an ABC classification and are not subject to the turn ratio or cost models for those classifications (such as min/max). They may take on another classification such as D or other designator as directed by the SSC.

Figure 4-8 shows the key "first" steps to getting this practice moving.

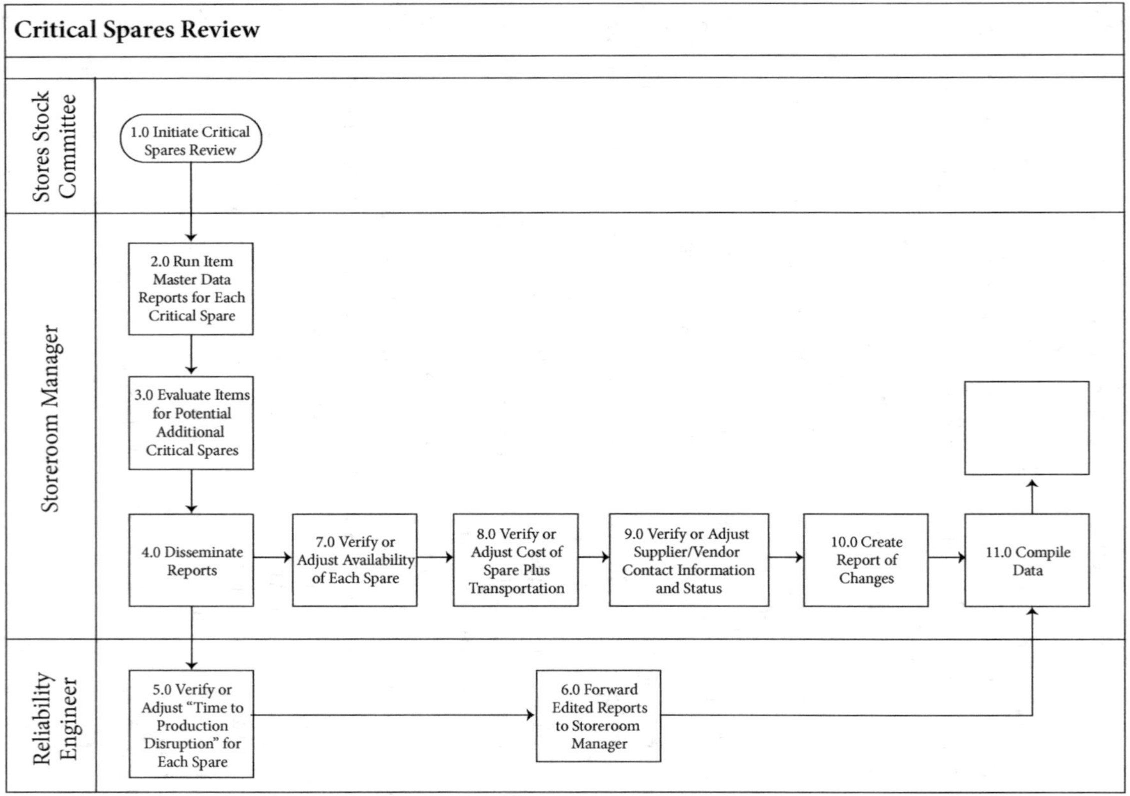

FIGURE 4-8 Critical Spares Review (partial).

A favorable outcome to the successful review of critical spare parts candidates will be that only a handful of parts rise up to this high-water mark. You only want a few critical spare parts. I would counsel that you want no more than 1–2% of your parts to be listed at this exclusively high level. For a storeroom with 1,200 MRO spare parts line items, that would be 24 critical spare parts, max.

These parts will have been vetted through a rigorous and auditable process. No longer will a part be labeled as "critical" just because someone said it was. The parts must meet challenging criteria. You don't want to have many critical spare parts. Believe me, you don't.

If you've taken one of my classes, it is a sure bet you've heard this next description of a critical spare part.

Imagine that you have two parts. One of the parts is out on the factory floor, attached to the asset, and performing its function. The second part is in the storeroom, sitting in inventory. The part in the storeroom is classified as "critical," and the storeroom associates are directed to make sure it is always ready to go, serviceable, and available at a moment's notice. For goodness' sake, it's critical! The plant will shut down instantly without this part in inventory.

Now let's walk out to the machine where the other part is working away. When we get to this part, we will likely find it covered in work residue. Physical evidence would indicate that it hasn't been cleaned or serviced in weeks. There is likely to be an operator sitting his cup of coffee on it, and a mechanic hitting it with a hammer. Heck, this part, this critical part, doesn't even look to be that *important* out on the floor. How can it be critical in the storeroom and of no apparent consequence while in use?

I tell my classes that when I'm approaching the in-service component whose companion is stocked as critical in the storeroom when I first see this item, I should be forced to avert my eyes due to the sheen of "cleanliness" emanating from the object. It is the cleanest object in the facility. As I approach the part on the machine, I feel a softness under my feet. I look down. It's a prayer rug, laid out in homage to the part. What's that, I ask? Incense burning to the left, a possible animal sacrifice to the right! That's how important this critical spare part is to my operation. If you don't have that, or something approaching that kind of care and interest, are your parts even that critical?

For a part to be critical, it should be treated as such. The part in stock will be under a separate level of security. There is no reason for anyone to be messing with a critical spare part in storage. In the example storeroom layout map shown in Figure 2-2, that company had intentions of stocking its critical spare parts on the mezzanine, way in the back.

For a part to be critical, the actual part that is being used in production needs to have the world's best preventive and predictive maintenance scheme aligned for its continued, valuable service to the organization. I tell my storeroom clients to check and see if their colleagues in the maintenance department have the best-of-the-best strategies in practice for parts deemed to be critical. One criterion that separates a critical spare part from a regular spare part is that you never, ever want to use it. More on that in the next zone.

For now, remember that the Critical Spares Review is the practice of evaluating those candidate parts the SSC believes to be uber-important against the criteria the SSC has determined to classify a part as critical. This review is accomplished here in the Intermediate-Effective zone.

Let's work to put the "critical" back into critical spares.

What this practice and the process of ABC Classification Review set us up for is the next stop on our drive through the Intermediate-Effective zone . . . Cycle Counting.

CYCLE COUNTING

The Cycle Counting process is very easy to understand. Even the reason it is necessary is straightforward. What makes this element so frustrating is trying to fit it into the busy storeroom schedule and understanding the totality of the value for the effort.

Cycle counting is the leading indicator that feeds into the lagging indicator (and metric), Inventory Accuracy. Breaking down the term, "cycle counting" is *counting* on a *cycle*. That sounded more sophisticated in my head. Sorry about that.

Cycle counting is simply the act of counting what is physically in inventory and comparing it against the count that is shown in the inventory management system.

The *count* should be random and should be a blind count. "Random" indicates that you don't know what items will be counted or when. "Blind" indicates that the person performing the inventory count doesn't know what the count should be. This practice provides an unbiased output from the audit.

A mistaken belief exists that the more we count something, the more accurate it is going to be. I've always found this to be an interesting part of human nature. There is a response most humans have that forms a perfect analogy. Everyone reading this book has, at one time, looked in the pantry or the refrigerator for something to eat. Finding nothing in the pantry or fridge, we will return to the same location and look again! It is almost as if we are expecting an overlooked plate of cold fried chicken or a box of leftover pizza to suddenly come into view. It's not there.

Counting something or looking for something over and over again doesn't make it appear. The purpose of cycle counting isn't directly connected to an accurate inventory. It is, however, linked. Let me explain.

Most storeroom associates would be quick to blame an inaccurate inventory or missing stocked items on one cause—maintenance technicians are removing items from the storeroom without recording the removal. If you have the unenviable situation of maintenance technicians being required to find a part in the storeroom themselves, you are probably constantly having a battle with them to sign the guestbook on the way out. Is that too much to ask?

Technicians failing to record parts removals without documenting the action is only one reason an inventory count can be off. The storeroom itself can cause the inventory count to be off for several reasons. A list of those activities includes, but is not limited to:

- Receiving parts incorrectly ("each" instead of a box)
- Putting parts on the shelf or job kits incorrectly or in the wrong spot
- Kitting incorrectly
- Cleaning and other 5S activities without paying attention
- Repair or replace process executed incorrectly
- General carelessness
- Performing PM activities on parts incorrectly, or paying less than focused attention
- Issuing incorrectly (an entire box instead of "each")
- Delay in recording goods movements
- Obsoleting a part, but not removing it
- Mishandling of damaged or quarantined parts

It is true that when maintenance technicians fail to record the removal of a part, it almost always leads to an inventory discrepancy. But look at all the opportunities the storeroom personnel have to mess up the inventory count.

Properly (or poorly) executing the activities just discussed *leads* to an accurate (or inaccurate) inventory count. Cycle counting is the leading input to having an accurate inventory. The results of cycle counting tell you if you are performing all the storeroom functions properly.

The recommendation is to cycle count every day. You aren't required to count the full inventory every day, just a little at a time. You should, however, have a process by which the entire inventory is counted at least twice a year. Now you can appreciate that storerooms should ideally have a low inventory volume and work to have every required part order directly in for a job kit.

Remember, cycle counts are to be random and blind. How do you pull this off?

Table 4-3 is an example of what a daily, random, and blind cycle count might look like. This is for a moderately sized storeroom that has 1,500 MRO line items. For this exercise, only concern yourself with cycle counting maintenance spare parts. This would include parts and materials for maintenance activities. For daily activities, it is common to calculate monthly activities using a standard 20 days per month, as it isn't recommended to ever use overtime to complete such tasks (e.g., weekends, holidays, personal time off).

TABLE 4-3 Cycle Counting Example

Classification	% of Inventory Volume	# of Line Items	Count Every:	# of Days in Count Period	# to Count Every Day
A items	15	225	Month	20	11
B items	35	525	Quarter	60	9
C items	48	720	Semiannual	120	6
Critical spares	2	30	Week	5	6
				Daily count	32

From Table 4-3, you can determine some very useful information. If you count A items monthly, B quarterly, and C semiannually, you will have counted the entire inventory at least twice in a year's time. In this example, D items represent critical spare parts, and through the Critical Spare Parts

Algorithm practice (yet to be discussed), the critical spare line-item count is held to a manageable 2% of the inventory.

From the table, if you count a little A, a little B, and so on each day, you would be required to cycle count 32 items a day. That seems very manageable for a storeroom with 1,500 MRO line items. *Note:* You never count the same item twice in the cycle of its counting. For example, if you have 225 A items to count in a month's time, you only count each item once. Those A items do not come up for recount until the next month.

The Stores Stock Committee is the entity that establishes the daily count requirement as well as establishes the tolerance that will be accepted during the counts. Developing the criteria is accomplished in the Cycle Counting Criteria process, which is in the Intermediate-Efficient zone to be discussed later.

I had remarked earlier that some companies have a difficult time fitting cycle counting into their storeroom schedules; this is particularly challenging when there is only one person working in the storeroom. I've further mentioned, many times, that there are no good reasons for a technician to be in the storeroom. Let me give you the exception to the rule.

For storerooms with limited personnel, there is a wonderfully simple solution to accomplishing the cycle count, plus other activities as needed. The exception to the rule is to have a technician with a work order perform the cycle count. How would this work?

As normal, at the beginning of the technician's shift, the technician is given a work order. The technician's assigned task for the next hour is to report to the storeroom. In the storeroom, the technician is given a list of 32 parts (in our example. . . a little A and little B, etc.). The technician has the part number, description, and location. Randomly selected, and without the actual anticipated count. Random and blind. The technician makes the count and reports the findings. It is that simple.

There are other positive results of having a technician perform the cycle count. The technician is likely to be familiar with the part and able to identify a part that is incomplete and/or no longer serviceable. The technician can also readily tell if a part is out of place or stored incorrectly. The technician may also grow a greater appreciation for the value of an accurate inventory and make an effort to be a better partner with the storeroom. Remember that a technician or two are members of the Stores Stock Committee.

If set up and executed correctly, this is a very positive process.

When the cycle count is complete, the inventory must be adjusted to match what is on the shelf. Efforts should be made to find and/or relocate parts to correct any discrepancies. This might include performing a root cause analysis if a trend is discovered. (RCA will also be discussed later.)

Some tolerance is allowed for counts that are off. The Stores Stock Committee determines these tolerances. Missing inventory may ultimately require a write-off.

Inventory accuracy needs to be at or above 98%. Care should be taken to ensure that A item counts are 100% accurate, as are critical spare parts. For some items, like caulk, tape, small seals, and bearings, you aren't going to close the ports if the count is off. However, if you are missing a 50-hp motor, you probably have some explaining to do.

Figure 4-9 illustrates a good starting workflow for the Cycle Counting process.

The Intermediate-Effective zone includes two very important metrics, or KPIs: Inventory Accuracy and Inventory Service Level. These two and one other work together to form the backbone of storeroom measurements. Though they are both designated as "KPIs" in the table of storeroom processes and practices (Table 1 in the introductory chapter, "The Journey Begins"), they are, however, quite independent and will be discussed in some detail.

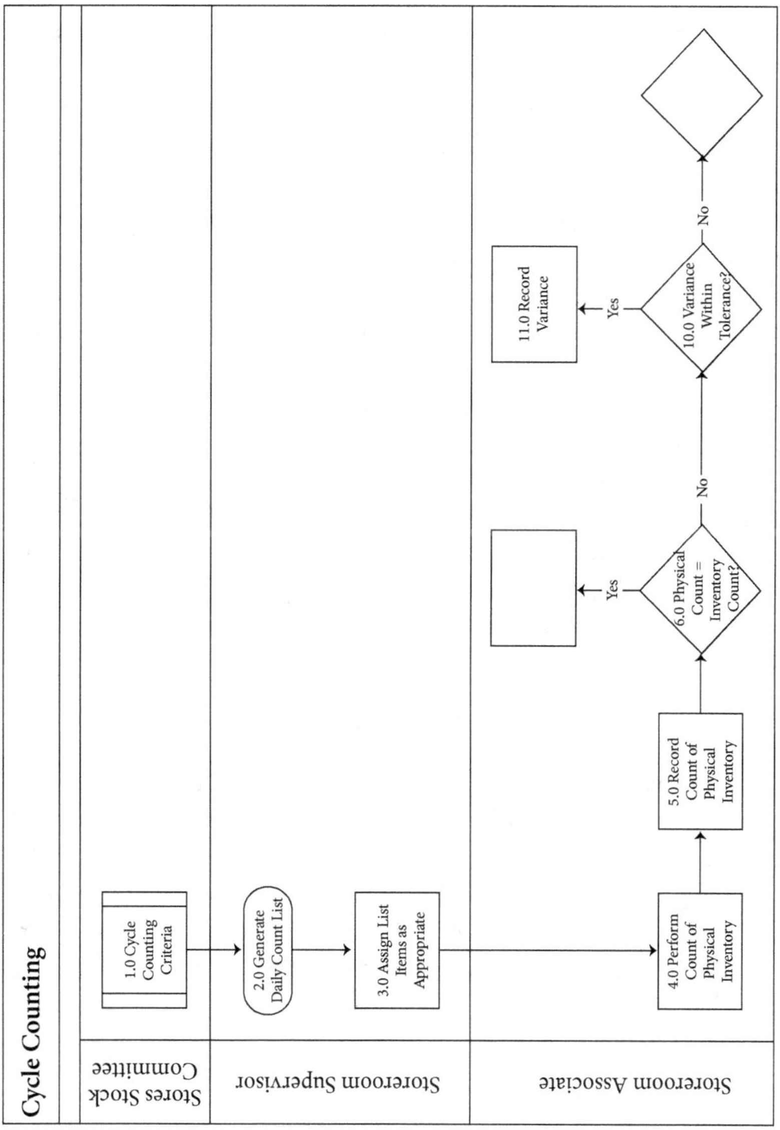

FIGURE 4-9 Cycle Counting (partial).

INVENTORY ACCURACY

We are arriving at our first metric, or key performance indicator, on our roadmap. If you recall earlier, we discussed the difference between a metric and a KPI (Figure 2 in the introductory chapter, "The Journey Begins"). If your organization has a stated goal or objective to enhance and improve these metrics and has taken formal steps to ensure efforts toward goal achievement, these metrics will be KPIs. I suggest that you definitely track these metrics.

Measuring inventory accuracy was just mentioned as part of the breakdown of cycle counting. Interestingly, the Society for Maintenance & Reliability Professionals is quiet on this metric, and I'm not sure why. It could be that there are no universally agreed-upon target values as they relate to best practices for this metric.

Ramesh Gulati provides some sound advice: "Because inventory accuracy is very critical for the maintenance store, it is important that a process is established to ensure that high inventory accuracy of 95% or better is maintained" (p. 136).

I counsel my clients to aim for an inventory accuracy of around 98%. Please take into account that my number and Ramesh's number are aggregated across the different classifications of parts.

For critical spare parts, there is no acceptable tolerance. If we are missing a 100-hp motor . . . close the ports! For items that are high value, low tolerance. For items that frequently turn, there is a very tight tolerance. Consider the stockout potential for an item with a high turn ratio that we control very loosely in terms of inventory accuracy. That will bite you with a stockout, every time.

I'm hoping that Table 4-4 will demonstrate the nature of the "running count," the practice of keeping track of inventory accuracy. I'm using some numbers that we developed for cycle counting. I thought it would be useful if we kept that scenario going.

TABLE 4-4 Inventory Accuracy Example

Day	Line Items Counted	# of Counts That Were Correct	Math	Daily Accuracy	Running Accuracy Total
1	32	31	31/32 × 100 = 96.87%	96.87%	96.87%
2	32	25	25/32 × 100 = 78.81%	78.81%	90.32%
3	32	27	27/32 × 100 = 84.44%	84.44%	86.46%
4	32	30	30/32 × 100 = 93.75%	93.75%	88.28%
5	32	29	29/32 × 100 = 90.62%	90.62%	88.75%

Table 4-4 shows that the inventory accuracy after Day 1 is 96.87%. Day 2's accuracy is 78.81%, but when added to Day 1's results, the average of the two days, or the running count for the accuracy, is 90.32% and so on.

What do we do with this information? Great question.

The data should be analyzed for trends. Is it the same category of item that has an inventory discrepancy (motors, gearboxes, bearings)? Is it the same item (bearing, seal, etc.)? How does January of this year compare with January of years past? Are we at least consistent?

Reliability Web published the results of an MRO Best Practices survey in 2017 with some numbers that seem to be concerning. The survey report includes a table, Table 1, "Guide to Performance Indicators for MRO Storerooms." One category was "Variance Report by SKU." The description of the results indicates that "< 2–3% Major company [sic] initiated a storeroom audit when [this] level hit 3%" (*Maintenance, Repair and Operations (MRO) Best Practices*, p. 15).

I read those results as "less than 2 or 3% of companies are concerned enough about inventory accuracy to initiate a storeroom-wide audit if their cycle count is off by as much as 3%." You might interpret the survey results differently, but this is troubling.

Inventory accuracy is a lagging indicator. In the preceding section, I discussed the leading activities used to move this lagging indicator value. Any time we "touch," or process, an inventory item, we create an opportunity to get it wrong. For all the tasks that require touching or accounting for an item, be sure to execute those with the highest level of competency and the highest level of consistency.

Very poor inventory accuracy can lead to many negative outcomes for the storeroom and those it services. Chief among those are:

- Stockouts
- Ordering parts that aren't needed, leading to excessive parts
- Wasted time looking for something that isn't there
- Issuing the wrong part due to incorrect stocking (this can be very dangerous)
- Frustration and low morale across the facility

There are a couple of key concepts for attaining inventory accuracy:

- Cycle counting is the primary input toward inventory accuracy.
- This metric can be kept at a high success percentage by accurate receiving, issuing, and storeroom security.

Inventory Accuracy is a metric that I call one of the "Sergeant Majors." If you have a military background, you might recall that a Sergeant Major in the Army and the Marine Corps is an E-9, the highest of the enlisted ranks. I have a few select metrics in maintenance and reliability that I always refer to as the "Sergeant Majors."

Another Sergeant Major is Inventory Service Level.

INVENTORY SERVICE LEVEL

If you look at the roadmap, you'll notice that one way out of the Intermediate-Effective zone is through the metrics. The Inventory Service Level and its companion Document Stockouts are quite possibly the two elements that I get asked about the most. I think that is because there are many sources offering somewhat conflicting advice on what service levels and stockouts mean and how to calculate them.

Not one to be shy about offering my opinion, especially in a book that I am writing, I will tell you that this metric is simply the number of times someone requests a part or material from the storeroom *divided into* the number of times the person is satisfied with the parts/material issued (x 100). How can that possibly be confusing?

In their *Glossary of Reliability Terms and Definitions*, Ho and Hodkiewicz present the clearest understanding of service levels: "The defined service quality for a particular activity or service area against which service performance may be measured. Service levels usually relate to quality, quantity, reliability, responsiveness, environmental acceptability and cost" (p. 12).

Ramesh Gulati focuses on the concept of storerooms by defining a service level as an "inventory level at which demand for an item can be met from the on-hand stock. Usually, expressed as *percentage of order satisfied*" (p. 146).

Here is the confusing part. At the beginning of this section, I said that a service level is the number of times someone *requests* a part or material from the storeroom divided into the number of times the person is satisfied with the parts/material issued (x 100). It would seem more correct to define the service level metric as the number of times someone *needs* a part or material from the storeroom divided into the number of times the person is satisfied with the parts/material issued (x 100). Did you see what I did there? If not, read it again. The first definition I had was "requests," and for the second definition, I changed it to "needs." Wouldn't you rather have a storeroom that can satisfy all the *needs* vs. one that strives to satisfy all the *requests*? Those would be two completely different storerooms. The challenge is that you must define what a *need* means at your location.

To compound this issue, storerooms should only be held accountable for the items they were requested to stock. Do you recall the very first process element, New Item Setup? If the item was never formally set up, the storeroom can't, or at least shouldn't, be held liable for having it. In fact, the storeroom shouldn't even have a part that wasn't formally set up. That makes sense. Right? This tiny detail will play a big part in the Document Stockouts element.

Storerooms should strive for an inventory service level of 97% or greater. Like the inventory accuracy metric, this is calculated as a running total. Table 4-5 provides an example of how this might be calculated.

TABLE 4-5 Inventory Service Level Example

Day	# of Requests	# Filled Completely	Math	Daily Service Level	Running Service Level Total
1	25	25	25/25 × 100 = 100%	100.00%	100.00%
2	31	24	24/31 × 100 = 77.42%	77.42%	87.50%
3	14	13	13/14 × 100 = 92.83%	92.83%	88.57%
4	30	30	30/30 × 100 = 100%	100.00%	92.00%
5	18	17	17/18 × 100 = 94.44%	94.44%	92.37%

What do you do with this information? Does that sound like a familiar question?

The data should be analyzed for trends as we did in the Inventory Accuracy element. Does a low service level measure the result of items requested from the same category? Is the failure to fulfill a request always suffered by the same person, same shift, or same circumstances? Is it the same item (bearing, seal, etc.)? Is not stocking the part the issue, or is it having the correct *quantity* that seems to hound you? How does January of this year compare with January of the last few years? Are you at least consistent?

The absolute best way to raise a storeroom's metric on the service level is to plan and schedule maintenance work and to kit parts. Reducing the demand for parts and materials at the storeroom service window will improve service. It is not the subject of this book, but there are many references to the best practice of waiting to schedule a maintenance job until all the parts are on-site.

As a continuation of that thought, the best planning and scheduling is only possible with exceptional preventive and predictive maintenance. PM/PdM begets planning and scheduling, which begets a reduction in urgent work, which begets an increase in planned work, which begets an increase in kitted parts, which begets an increase in inventory service level, which we will discover begets a reduction in stockouts. And finally, better maintenance budget control.

DOCUMENT STOCKOUTS

We made it to the end of the Intermediate-Effective zone. This last element could have been labeled simply as "Stockouts," but by including the word "Document," I had hoped to stress the importance of recording, analyzing, and improving the results of this process. The Document Stockouts process appeared as a metric on the original Storeroom Roadmap. I elevated it to a tactical process in recent years. This is an activity that has the potential to be executed every single day.

Stockouts are thoroughly discussed in the *SMRP Best Practices*. SMRP is often the source of great answers and my go-to source, but a grain of salt and a little research are needed to use this valuable guide effectively.

SMRP Best Practices defines stockouts as "the measure of the frequency that a customer goes to the storeroom inventory system and cannot immediately obtain the part needed" (p. 264). When this definition from SMRP is used in tandem with Gulati's definition of percentage of stockouts, our focus gets sharper. Gulati defines the percentage of stockouts as the "number of stockouts/total items issued" (p. 146).

Why is this an important activity to document and analyze? Because it "is used to maintain the appropriate balance in stocked inventory. Too much inventory increases working capital unnecessarily. Too little inventory results in unnecessary delay and equipment downtime that can negatively impact costs and profits" (*SMRP Best Practices*, p. 264).

SMRP suggests a best-in-class target value of "less than (<) 2%" (*SMRP Best Practices*, p. 266).

We begin by comparing SMRP's formula (p. 264) for calculating the percentage of stockouts with that previously shown as Ramesh Gulati's formula:

SMRP stockouts (%) = (number of inventory requests with stockout/total number of inventory requests) × 100

Gulati stockouts (%) = (number of stockouts/total items issued) × 100

I admit that these two formulas seem similar, but I can't shake the differences. I have done a great bit of study on this and have concluded that SMRP's formula is helpful when determining the percentage of stockouts over the course of time, where Gulati's seems to be helpful at the moment of issue.

Here is an example using Gulati's formula:

A mechanic requests two pillow block bearings; these are the same bearings (same line item). The storeroom only has one of those bearings.

Number of stockouts **(1)**/total items issued **(1)** × 100 = **100%**

That is a 100% stockout situations because the storeroom did not fill the request for a stocked item in the quantity needed.

Same scenario, but now the storeroom has the two bearings.

Number of stockouts **(0)**/total items issued **(2)** × 100 = **0%**

There is no stockout here.

Where this gets confusing can be illustrated in this next scenario:

A mechanic requests three pillow block bearings; these are the same bearings (same line item). The storeroom only has one bearing.

Number of stockouts **(1 or is it 2)**/total items issued **(2)** × 100

There are two schools of thought within the maintenance and reliability community. One school says that the storeroom should only be assessed one stockout since it was the same line item. The other school of thought indicates that the storeroom should take two hits because it shorted the customer by two bearings. What is your company's position on this? I would advise that you assess the storeroom only one stockout.

Another tricky interpretation occurs when the need is for 15 of the same item, and the storeroom only has 10. Is the stockout for one (since it is one line item) or for the five items not filled? When a part request is made for multiple parts of the same component, and the storeroom does not have enough inventory to fill that quantity, the storeroom is only assessed a stockout of one. Otherwise, requesting 20 of a particular item and only having five in inventory would result in a stockout of 15. That seems unreasonable and against the spirit of the measure.

Look at a more complex scenario and determine how you would apply Gulati's formula for stockouts:

A mechanic is working on a routine maintenance job. This job was not planned and scheduled, and the parts certainly were not kitted. After troubleshooting the asset malfunction, the technician goes to the storeroom with a valid work order and requests three line items. The mechanic needs two bearings (these are the same pillow block bearings), two sprockets (the same sprockets), and two feet of #60 diamond roller chain (unit of issue is per foot).

The mechanic makes the request using the part numbers for each item. The storeroom clerk fills the request for the two sprockets and the two feet of roller chain but only has one of the pillow block bearings.

How would you calculate the stockout situation in this scenario using Gulati's formula? Here is my first attempt:

Number of stockouts **(1)**/total items issued **(5)** × 100 = **20%** stockout rate

Note: In Gulati's formula, five items were issued: one bearing, two sprockets, and two units of chain.

However, and you just knew there would be a "however," *SMRP Best Practices* offers this decoding key under the heading of "CAUTION": "Stock outs are measured at the individual line items level and for total quantity requested/required versus supplied (e.g., no partial credit)" (p. 266).

Using this guidance from SMRP to specifically look at each line-item request separately, I am modifying my thinking. For the more complex scenario above, the stockout percentage for the sprockets and the chain would be 0% since the request and the fill are 100%. And for the bearing, it is a 100% stockout condition:

Number of stockouts **(1)**/total items issued **(1)** × 100 = **100%**

Does that seem like the logical conclusion to the scenario?

Let's focus on SMRP's formula, the one that I conclude is for an overall look at stockouts over a period of time:

> **SMRP stockouts (%)** = (number of inventory requests with stockout/total number of inventory requests) × 100

Use the following scenario for SMRP's formula.

A storeroom receives 1,234 stock requests in a given month. There were 30 requests in this same month where there was insufficient inventory to fill the request. What is the stockout percentage?

> Number of inventory requests with stockout **(30)**/total number of inventory requests **(1,234)** × 100 = **2.43%**

I confess that I might be reading and interpreting these two formulas wrong, but I can't shake the fact that SMRP and Gulati differ slightly. What is universally agreed upon is the need to document and analyze the situations where the storeroom cannot fulfill the need. This is the Document Stockout process.

You should always be documenting stockout occurrences. As rudimentary as this sounds, the practice of documenting stockouts is often just making notations on a clipboard hanging from the service window in the storeroom. Some very good storerooms where I've consulted record each "potential" stockout on a separate form and hold those for future investigation. I put the word "potential" in semi–air quotes to imply the softness of a stockout at the moment of inquiry. If someone asks for a part at the service window, and the storeroom doesn't have it, it isn't automatically a stockout. Why? Because there are many qualifiers for an actual stockout classification.

Here is a short, nonexhaustive list of some of those qualifiers:

- Is the part needed or just wanted?
- Is the part authorized with a work order?
- Is it a part that the storeroom was officially instructed to stock?
- Is the part normally stocked at the level (quantity) being requested?
- Was the last of that item issued with zero lead time to replenish? (Did the person in front of you take the last one?)

Another confusing and sometimes confounding occurrence when calculating stockouts is dealing with inventory demands that exceed what was agreed upon as the typical and usual usage.

SMRP anticipated a request for a quantity of an item in excess of its typical and reasonable consumption. The Best Practice standard provides two clear, instructive passages.

In defining the numerator for its formula, *SMRP Best Practices* explains the number of inventory requests with a stockout this way: "An inventory request is a stock out if the requested item is normally stocked on-site *and the inventory request is for a normal quantity of the item*, but the inventory on hand is insufficient to fill the request" (p. 264).

And buried in a sample calculation for stockouts is this very helpful example from *SMRP Best Practices*: "There were 30 requests in this same month where there was insufficient inventory to fill the request. Analyses of these 30 requests found four that were *excessive orders beyond normal request quantities; therefore, these four requests did not meet the criterion for stock outs*" (p. 265).

I don't know if I was completely successful in clearing up some of the confusion around stockouts, but I hope I did compel you to understand the importance of documenting stockout occurrences. I want to wrap up that point by informing you that your company should clearly define what a stockout means, how you will calculate a stockout, and, as important, what you will do with the information.

Aside from tracking and trending stockouts to determine if they occur on or around similar types of components, or the same maintenance or production shifts, it is also important to compare stockouts with the service level metric, cycle counting effectiveness, and the security around your storeroom's physical area of jurisdiction.

One of the more interesting tools I've used around the metric for stockouts is to overlay the stockout occurrences over a map of the storeroom to create an analysis tool called a "heat map." Figure 4-10 is one such example.

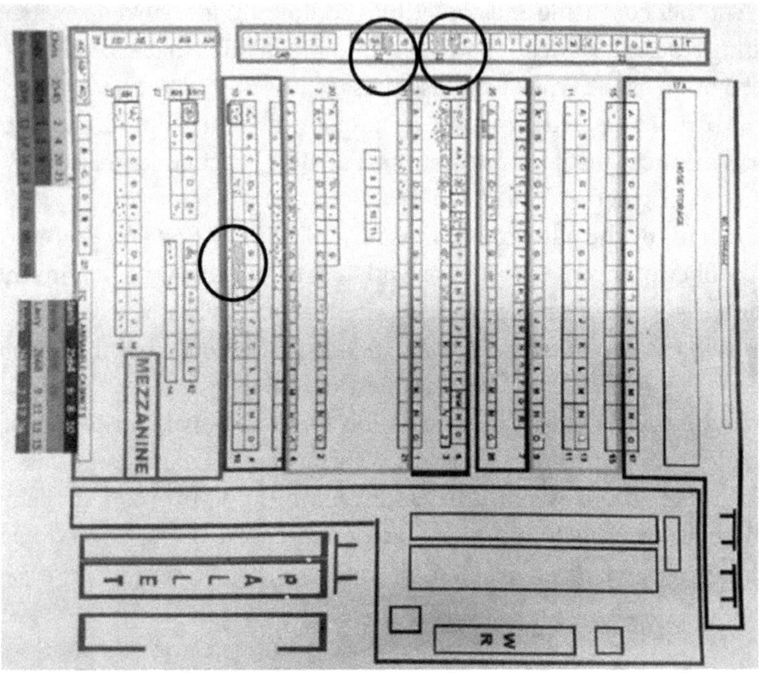

FIGURE 4-10 Heat map.

I concede that it is difficult to see what is going on with the heat map in a black-and-white photo. The color version is much more telling. The map is the plan view of a client's storeroom. Over that map, I made a small red dot at the geographical location of each documented stockout over the course of two years. The map has hundreds of red dots all over the place. But there are times when those dots are right on top of each other. I've drawn some dark circles around a few of those instances.

This heat map gave us a graphical representation of what was going on with the stockouts. The data originally came to us as a table of events that was very difficult to decipher. We converted that data into information via the heat map and could then make some assumptions that we later proved to be true. Our analysis showed that the client had two vendors that were underperforming. The client now had a table of data and a compelling graphic to make the argument for the vendors to shape up.

Document the definition of stockouts at your company and the formula you are going to use to calculate the value. Also, document how you are going to record a stockout. Some companies use a simple clipboard at the service window. Still others use elaborate slips of paper and drop boxes. Find a system that works for you. Being technically savvy is not as important as being accurate. Shoot for a value of less than 2%. Document the occurrences (and you will have many), and use the data to create the information needed to improve.

As with most maintenance and reliability metrics, stockouts will improve with greater preventive and predictive maintenance and more awesome planning and scheduling. I promise.

SUMMARY

The Intermediate-Effective zone is where I envision the Great Plains and our journey's foothills. We have established our discipline for purposeful execution in the Foundation-Effective and the Foundation-Efficient zones. In this new level of Storeroom Roadmap zones, we started some open field running and set into motion the activities that are going to sustain the momentum of our work. Forever.

We also introduced some nuance to our roadmap. There were three companion elements that were mentioned: the Adjust ABC Classification Model, Critical Spare Parts Algorithm, and Cycle Counting Criteria. These will make their mating processes and practice more efficient . . . when the time comes. We also got a glimpse of our first metrics, or key performance indicators. The difference between a KPI and other metrics is that a KPI is tied to a stated goal or objective. I recommend that you have these metrics operating and perhaps consider having stated goals to increase the metric value and make the measure a true KPI in your company. Why? Because the two discussed here so far are two of the Sergeant Majors.

As discussed in this zone, you must have a process to ensure the smooth return and possible refund of parts no longer needed for the original cause or those that might have been received in error. Essentially, how do you get a part back to where it belongs after it has exchanged hands?

We also thought about emergency or urgent purchasing and why those situations might exist. I know that as long as there are people driving fork trucks in your facilities, you will have emergency maintenance. As long as you have emergency maintenance, there will exist the need for emergency procurement. Always.

The storeroom has a charter and primary goal to have the right part at the right time and in the right quantity. To achieve those lofty goals, the inventory must have some order or stratification. Otherwise, a tube of caulk has the same intrinsic value as a 25-hp motor. And that wouldn't make sense.

We begin to determine the importance of one line item over another in the ABC Classification Review and the Critical Spares Review. To make sure everything is accounted for, you should get very good at cycle counting and understand how to execute this process consistently at the highest level.

WORKBOOK

Here is the assignment for Chapter 4. Fill in Table 4-6 with a Y for yes or an N for no.

TABLE 4-6 Answer Y or N

	Y or N	N/A
Do you have formal process guides created for the following?		
Return to Stock		
Return to Supplier		
Emergency Procurement		
ABC Classification Review		
Critical Spares Review		
Cycle Counting		
Document Stockouts		
Do you track inventory accuracy?		
Is your inventory accuracy > 98%?		
Do you track inventory service level?		
Is your inventory service level > 97%?		
Do you track stockouts?		
Are your stockouts < 2%?		
Do you perform blind and random cycle counts?		
Do you cycle count a little bit everyday?		
Do people have to answer for making emergency purchases?		
Do maintenance technicians bring back items for restock in their original containers?		
Are return-to-supplier items always in good shape and in their original containers?		
Are return-to-stock incidents at a minimum?		
Are technicians allowed to reshelf their return-to-stock items?		

I'm sure you'll find the short videos listed in Table 4-7 helpful and great conversation starters for your organization's understanding of the Intermediate-Effective zone of the Storeroom Roadmap.

TABLE 4-7 *Dr. John's Maintenance Minute* Links

Title	**Maintenance Innovators, Inc., Website Link**
Return to Stock	https://maintenanceinnovators.com/aiovg_videos/return-to-stock/
Return to Supplier	https://maintenanceinnovators.com/aiovg_videos/return-to-supplier/
Emergency Procurement	https://maintenanceinnovators.com/aiovg_videos/emergency-procurement/
ABC Classification	https://maintenanceinnovators.com/aiovg_videos/abc-classification/
Critical Spare Parts	https://maintenanceinnovators.com/aiovg_videos/critical-spare-parts/
Cycle Counting and Cycle Counting Criteria	https://maintenanceinnovators.com/aiovg_videos/cycle-counting-and-cycle-counting-criteria/
KPIs and Metrics	https://maintenanceinnovators.com/aiovg_videos/kpis-and-metrics/
Document Stockouts	https://maintenanceinnovators.com/aiovg_videos/document-stockout/
	YouTube Dr. John's Maintenance Minute Link
Return to Stock	https://www.youtube.com/watch?v=jaGlKBc7Fqo
Return to Supplier	https://www.youtube.com/watch?v=0J8IKK5WD4Q
Emergency Procurement	https://www.youtube.com/watch?v=g7SxkpTwFbo
ABC Classification	https://www.youtube.com/watch?v=OUKvbigzW-o&t=6s
Critical Spare Parts	https://www.youtube.com/watch?v=RkaZ9Ftj8ZI&t=8s
Cycle Counting and Cycle Counting Criteria	https://www.youtube.com/watch?v=ZMliel2fFPY
KPIs and Metrics	https://www.youtube.com/watch?v=XYHI7huFLLk&t=11s
Document Stockouts	https://www.youtube.com/watch?v=bzRPtfOVz0o&t=30s

CHAPTER 5

INTERMEDIATE-EFFICIENT

The two most efficient ways to attain maintenance value are PM optimization (to free up valuable resources) and reduction of spare parts (to save money).

—Dr. Klaus M. Blache

Making the analogy that the Intermediate-Effective zone of our roadmap was like the Great Plains and foothills we might encounter on a trip, it wouldn't be a stretch to liken this Intermediate-Efficient zone to the top of the mountains with a graceful slope to the green valley that is our destination.

The Intermediate-Efficient zone will have a little bit of everything. We will encounter both processes and practices, as well as organizational elements, and look toward managing the value proposition that is promised in an effective and efficient storeroom operation.

Figure 5-1 lays out the course that has us winding up the mountains and down the other side. Table 5-1 is the corresponding legend.

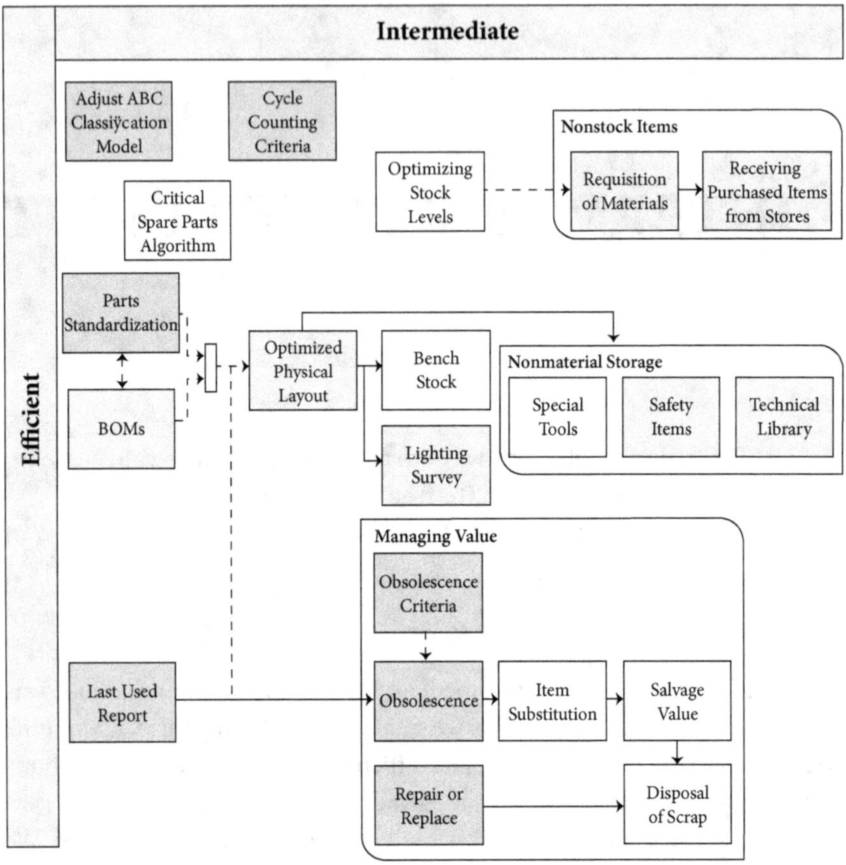

FIGURE 5-1 Intermediate-Efficient zone.

TABLE 5-1 Legend for the Intermediate-Efficient Zone

Element	**Activity**
Adjust ABC Classification Model	Process
Critical Spare Parts Algorithm	Practice
Cycle Counting Criteria	Process
Optimizing Stock Levels	Practice
Requisition of Materials	Organizational
Receiving Purchased Items from Stores	Organizational
Parts Standardization	Process
BOM	Organizational
Optimized Physical Layout	Organizational
Special Tools	Practice

Element	**Activity**
Safety Items	Organizational
Technical Library	Organizational
Bench Stock	Practice
Lighting Survey	Organizational
Last Used Report	Process
Obsolescence Criteria	Process
Obsolescence	Process
Item Substitution	Practice
Repair or Replace	Process
Salvage Value	Practice
Disposal of Scrap	Practice

We will begin traveling through this area of the map by navigating through the companion elements we encountered in the previous zone and had parked off to the side of the road.

ADJUST ABC CLASSIFICATION MODEL

The Adjust ABC Classification Model is the companion process to the ABC Classification Review. From Chapter 4, recall that the purpose of classifying spare parts as A, B, or C or some other means of stratification is to categorize groupings of parts into a family of components by some shared characteristic—cost, turn ratio, etc. Ramesh Gulati recommends that parts classified as A have a high value and make up a low volume of the inventory and so forth. By classifying parts into an ABC configuration, we set ourselves up perfectly for Cycle Counting, and for all matters as they relate to how much we care about the parts being available and on hand. As stated a few times, it does force us to look at a tube of caulk differently from a 25-hp motor.

This element in the Intermediate-Efficient zone is necessary to make our ABC review more efficient. Table 4-2 was a summary from Gulati's *Maintenance and Reliability Best Practices*. Table 4-2 is reprinted here to remind you of the range for A, B, and C items.

TABLE 4-2 ABC Distribution Spread (Reprinted from Chapter 4)

Category	% of Total Inventory Value	% of Total Inventory Volume	Average % of Volume
A items	60–80%	10–20%	15%
B Items	15–25%	20–35%	25%
C Items	5–15%	55–75%	65%

Using Gulati's numbers gives your Stores Stock Committee a starting point to consider the acceptable range at your location. As with other important and far-reaching decisions, the counsel of the SSC should be considered for this process.

For example, in the process guide (do you remember process guides from Chapter 1?), the SSC may determine that the spread, or stratification, for spare parts classification will be aligned with what is shown in Table 5-2.

TABLE 5-2 SSC's ABC Distribution Mandate

Classification	% of Total Inventory Volume	Plus or Minus	Range of Inventory Percentage
A items	15%	2%	13–17%
B Items	35%	2%	33–37%
C Items	50%	2%	48–52%
Critical Spares	2%	No more than 2%	0–2%

With this direction, the review can't be accomplished until the category percentages are met within the tolerance allowed. Figure 5-2 is a good example of what the adjustment workflow might look like. Pay particular attention to activity blocks 8.0 and 9.0. In these activities, the algorithm for calculating the values for the A, B, and C items is adjusted.

Remember, there are many ways to stratify your inventory. If you have a major brand of CMMS with an inventory management system, it is likely that your computer can perform this calculation for you. If so, it is almost certainly through an algorithm that considers the value of the component and the frequency of usage of that item. These two measures are also the suggested means of calculation by Gulati.

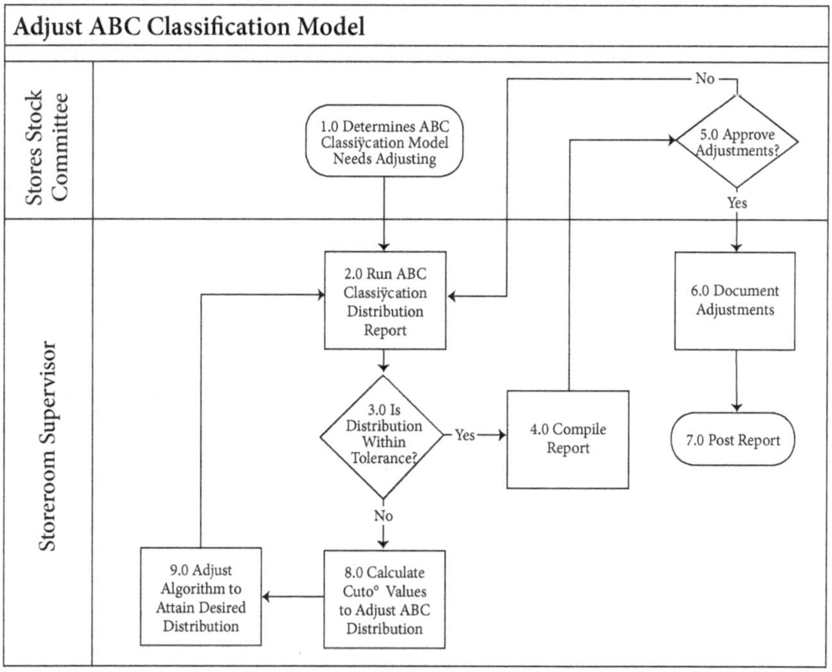

FIGURE 5-2 Adjust ABC Classification Model (partial).

It is not uncommon, however, for companies to base their ABC classification on frequency of usage alone. It is easier, and you don't need to be a math wizard to figure it out. I will conclude that point by expressing an interest in appealing to the financial side of your company. The financial folks are most likely going to be interested in the inventory value and the cost of the parts, whereas maintenance is understandably interested in having the parts on hand when needed. It never hurts to *play nice* with the folks in finance.

CRITICAL SPARE PARTS ALGORITHM

The Critical Spare Parts Algorithm is the companion element to the Intermediate-Effective zone element of Critical Spares Review. The review, as explained in Chapter 4, is to determine if a spare part's measurable characteristics raise it to the level of being critical. If there is not a repeatable, definable, measurable, and consistent method to determine if a part is critical, then anything can be critical. It would come down to a spare part being proclaimed as "critical" because someone said it was. Does that sound like a world-class process?

The Critical Spare Parts Algorithm determines guidelines for classifying spare parts as critical items. Using the algorithm, parts are assigned an output value. If this output value does not meet or exceed the cut-off criteria, the spare part is not automatically classified as critical. Parts may be reclassified later.

In Chapter 3, I mentioned that the Stores Stock Committee was going to prove to be very important and valuable. Here again, in Chapter 5, we see that value. The SSC creates the algorithm to determine a part's criticality. In all cases, our ability to classify, stratify, and stock at min/max levels must be based on math and measurable characteristics. At some point in our journey to build the best storeroom, we have to stop *guessing* at the numbers.

If you or anyone else were to be asked to state what characteristics make a part a critical part, that list would likely include:

- Not having the part shuts down production.
- The part has a long lead time.
- The part is expensive.

Without exception, these are some very important characteristics. Some questions are necessary to answer to properly understand the implication of the characteristic:

- Not having the part shuts down production, but does that mean all production or part of production?
- Does the part have a long lead time? If you must hold your breath until you get the part, 26 seconds is a long time.
- The part is expensive, but how expensive? A $100,000 part can be absorbed by a big company, but it could bankrupt a mom-and-pop operation.

The SSC will provide the answers to these questions. It is in its charter. How much production will be lost? How long is a long lead time? And how expensive is expensive? The SSC will address these and many other factors that make a part a critical spare part.

What follows is the discussion of a critical spares calculation form. With this form, the Stores Stock Committee and other associates will be asked to take a hard look and consider the hard truth. In each category of inquiry, the group must answer as to the impact that will take place within several specified categories. They (the SSC and others) must arrive at a value. The value is calculated as factor × weight to arrive at a sum. In the end, the individual sums are tallied, and a final result is given. The Stores Stock Committee determines what final score is needed to make a part "critical."

Let me share an example of this calculation form, or what I refer to as the "algorithm," in Figure 5-3.

CATEGORY	EXPLANATION	FACTOR	WEIGHT	SCORE
Time to Production Disruption: Time till plant shuts down?	Immediate	5	5	
	< 1 Day	4	5	
	< Week/50% Cap	2	5	
	< Month	1	5	
	No Impact	–2	5	
Availability: Time from order placed to arrival?	˜ 3 Months	8	5	
	1–3 Months	6	5	
	< Month	3	5	
	< Week	1	5	
	< Day	–2	5	
Cost: What is the cost for delivery?	> $10,001	5	1	
	$1,001–$10,000	3	1	
	$500–$1,000	2	1	
	< $500	1	1	
Supplier: Reliability of supplier based on rating?	Certiÿed	1	1	
	No History	3	1	
	Poor Rating	5	1	
	Consignment	–20	1	
Identical Equipment: How many duplicate pieces of equipment use the same item?	Multiply Total Assets x 4			
Obsolete: Will the part be available to buy throughout the lifespan of the equipment?	Yes	0	1	
	Not Sure	10	1	
	No	5	1	
Inspection: Do inspections always provide early warning?	Yes	–1	1	
	No	2	1	
Failures: Number of replacements over the last two years?	Number of Replacements x 5			
Shared: Stocked at other locations close enough to ship economically?	Yes	–4	3	
	No	5	3	
	Don't Know	1	3	
TOTAL POINTS ACCUMULATED:				

FIGURE 5-3 Critical spares algorithm.

Let's break these categories down.

Time to Production Disruption. This has traditionally been the number-one qualifier for a part to be categorized as critical. Notice the column titled "Explanation"—it is here that the committee is forced to consider options: immediate stop, less than a day until stoppage, less than a week *or* an immediate drop to 50% capacity, and so on. The SSC finds the most likely result of *not having the part* and then moves across to note the corresponding factor and weight. The multiple of factor x weight is recorded in the Score column for that category.

Note: The values and explanation are editable for each factor and weight—you can change them to any value you'd like. This is also true for the entire document.

Availability. This category is another traditional concern for critical parts. The text mentions under the category title that availability is a measure of the time between placing an order for the item and receiving it. As with the Time to Production Disruption category, there are several options in the Explanation column and corresponding factors and weights for the SSC to consider when arriving at a final score. All this text is editable, as stated in the note above.

Cost. This is an interesting category. It makes sense that a part that is highly disruptive to production and takes a long time to get a replacement would be expensive. The debate in the maintenance and reliability community centers on how to calculate that cost. One group feels that the cost of the part itself should be included in this determination. The other group feels that only the cost of shipping or delivery is the factor to consider. I tend to side with the second group, which is clearly in the text within the category box.

Supplier. This category is one that most people forget or never think about. A "certified" supplier is a supplier your company has a contract with. The approved supplier has all its licensing and insurance documentation on file. A certified supplier knows you and your company and is poised to respond to your needs. When you call a certified supplier, someone will answer the phone.

Consider this scenario. You have a 100-hp motor sitting in your storeroom as a verified critical spare part. It has been evaluated using the algorithm, and as anticipated, the sum of all the category scores was exceptionally high. You also have a 100-hp motor in use in the facility as the production companion to the critical spare. One day the 100-hp motor in use fails. Probably spectacularly. The critical spare motor is called into service. What do you do with the failed motor? You will most likely need to send it out for evaluation and repair if possible.

Here is the "rub," as might be said. When you call your motor repair company, does someone answer the phone? Chances are, the employees already have work on the books and in their shop. Each one of those jobs has a promised delivery date. Is the vendor likely to stop everything in its revenue-generating business to come running to your aid? A certified supplier will. That's the deal we have set up with that supplier.

The supplier category is also the first category we come to in our calculation where there are measurement options for uncertainty. For example, in this category, we have an explanation of "no history."

In all measuring options, you must go with the worst-case scenario if the answer is not known or absolutely known. If we don't have a history with this supplier, our only reasonable response is "no history." That is a factor of 3 and a weight of 1, which gives us $3 \times 1 =$ for a total of 3.

Identical Equipment This is a simple concept. If you have a spare component that others feel is so important that it should be evaluated to see if it is a critical spare part, then that situation is exacerbated by having more than one unit in production. The risk factor has increased, and this increase should be considered.

Obsolete. This is a new category and one that takes some research. The SSC cannot just phone this in. Give some thought to this scenario.

Your organization has a major production asset that is vital for your company's success. On that asset is a particular component that is one of the key parts to ensuring the function of that asset. Doesn't it seem to you that your organization should know if that part is going to be available in the future? If not, why not? It seems that an army of engineers ought to know this and stay on top of it.

Again, in this category, there is an option for uncertainty. Use the worst-case scenario if the facts around the issue are uncertain.

Inspection. This is vital. Even a part that is being evaluated to determine if it rises to the value of being critical should already have the world's best preventive and predictive technologies and practices thrown at it. Be honest with this category. If your answer is no, you had better get to work. No excuses.

Failures. This is a category with a heavy penalty. It makes sense that if your organization has a very important component in use and it has experienced repeated failures there ought to be some penalty in the calculation to compensate for that. This is a measure of failures over a set period (I suggest two years). The number of failures (or replacements) is multiplied times a meaty value to net the sum.

Shared. This is a very delicate category. A part stored in a (often) central location, with the understanding that it is available for all facilities if needed, is known as a "strategic spare." In this category, the SSC is being asked to determine if this is the case. Can a part be shipped from another facility to fill a need, a critical need? Note the word "economically" in the text within the category box. That is a very subjective word. The SSC needs to be on its game in this category and be honest about the answers.

There is a worse-case offering in this category as well. Here is a word of caution to consider. If our engineers have done their job, and our company subscribes to the Total Productive Maintenance philosophy of *standardization of parts*, you can answer favorably to this inquiry. That's an awful lot of ifs. Also, does your sister plant know that it is responsible for coming to your rescue? Is there a formal agreement or corporate governance?

That's the Critical Spare Parts Algorithm. As I mentioned many times, everything in and on this form is editable. I strongly suggest that the Stores Stock Committee consider this tool and create values and categories that make sense to your organization.

Use the following scenario and the algorithm tool to see if you arrive at the same conclusion I did.

As part of its semiannual duties, the Stores Stock Committee of the Atlas Company in Tulsa, Oklahoma, is meeting to discuss a few stocked components and determine if they warrant consideration as critical spares. As part of the creation of the Critical Spare Parts Algorithm, the SSC determined that any part with a final total of 85 or greater would be considered a critical spare part. Otherwise, the company will still stock the part, but it just won't be handled as a critical spare.

The next part up for consideration is a hydraulic cylinder. The cylinder is 6 feet long and 18 inches in diameter and has a 4-foot stroke. The cylinder is used to index the turntable, one of the facility's most critical assets. Every single product transitions through that turntable. If this cylinder were to fail, the plant would be down immediately.

If the cylinder were to fail, a replacement or repair would take approximately 14 days. The cost to have one shipped from the manufacturer is $5,000. The OEM (original equipment manufacturer) for this component is a long time vendor, and Atlas has an exceptionally good and historically well-served history with the vendor. There is only one use for this cylinder in the facility. It is uncertain if this device will even be available for purchase in a few more years, even though Atlas intends to use this turntable for the next 25 years.

Preventive maintenance includes visual inspections by maintenance, look-listen-and-feel by system operators, and hydraulic fluid analysis. There have been minor issues in the past, but no failures in the last two years.

No other facility in the company has this hydraulic cylinder.

Table 5-3 shows my results for this scenario. How did you do in comparison?

TABLE 5-3 Critical Spares Algorithm Score

Category	Explanation	Factor	Weight	Score
Time to Production Disruption	Immediate	5	5	25
Availability	< Month	3	5	15
Cost	$10,000–$1,001	3	1	3
Supplier	Certified	1	1	1
Identical Equipment	Just the one	1	4	4
Obsolete	Not sure	10	1	10
Inspection	Yes	–1	1	–1
Failures	0 (2 yr history—no failures)			0
Shared	No	5	3	15
			Total Points Accumulated	72

In the scenario, I calculated a sum total of 72, which is well short of the SSC's requirement of 85. So this part is not critical. Not under the rules.

Keep in mind a few things that we already know. Just because a part is deemed to be "not critical" doesn't mean we will not stock the part. In fact, we already have the part in the storeroom. We don't stock it as critical. That means we aren't storing it in that uber-secure area with the other critical spares. We aren't going to demand that maintenance have the best preventive and predictive maintenance in the world. And we are not likely to insist that the part in production be the cleanest part on the hill. And we are not going to cycle count it weekly.

We will likely stock this cylinder as an A item. We will take great care of it in the storeroom, and it will be ready and available when needed.

We want as few critical spare parts as we can have.

Let me touch on the genius of this critical spares algorithm form. Now you know, without a doubt, the categories that make a part critical, and what to do to make a part *less* critical. Use that information to your advantage and the advantage of your storeroom.

You want very few critical spare parts. By working to gain the information in the critical spares algorithm, and to improve the score, you will be able to move items out of that top classification. They will still be important parts, and you will still always have them in stock.

I want to issue a very big *however* to my recommendation to classify spare parts as critical based on their evaluation using a repeatable and measurable calculation. A critical spare can also be anything the plant or site managers say it is. I know this flies in the face of the entire theme of this element, but sometimes it comes down to whose *neck* is on the line. There should be some perks to having the top job, and let's assume that the boss is under a lot of pressure and will ultimately make a sound decision.

There is another quick litmus test to spare parts. You cannot have a critical spare on an asset that is not critical. Have you ever considered this? The *child* cannot be more important than the *parent*. That is a very easy way to look at it.

Also in wrapping up this element, I'm embarrassed to admit this, but the idea of *insurance spares* didn't click in my mind until I read Doc Palmer's book. I had heard for years how critical spares are sometimes called "insurance spares." I assumed this was just a play on words, *insurance* meaning "we have a sense of ease regarding our risk exposure because we have a spare part in our back pocket." Metaphorically speaking.

Doc helped me put this in perspective: "Critical spares are not just connected to plant capacity but might be required for critical environmental, safety, or even legal reasons. Some critical spares might be required to be on hand to obtain better insurance rates, as dictated by risk-management personnel" (p. 343).

CYCLE COUNTING CRITERIA

This element is the companion element to the element of Cycle Counting, which was detailed in the Intermediate-Effective zone. As a reminder, "cycle counting" refers to the routine counting of on-hand inventory and comparing that count to the recorded amount in the inventory management system. Each day, a small number of items are physically counted and compared with the inventory count. This allows decisions and adjustments to be made if the counts do not coincide. By monitoring the accuracy of the inventory daily, actions that might compromise the inventory accuracy can be detected and corrected more quickly.

The Stores Stock Committee is responsible for establishing the criteria for cycle counting. The execution of inventory counts must be consistent. Because inventory control is critical, all inventory accuracy guidelines are aimed at maintaining unbiased and random counts (to avoid gaming the system). I've had clients that would divide up the storeroom into areas of responsibility, with each associate having a row or two to maintain. The day prior to their cycle count, these associates would spend some time straightening up the row they were going to cycle count the next day. Not surprisingly, they had a high inventory accuracy.

The SSC should establish some key concepts and communicate those to ensure that the stakeholders interested in the storeroom's performance know how the performance of their own divisions and departments can positively or negatively affect inventory accuracy. Some of those key concepts include the following:

- Any item received, tracked, and controlled by the storeroom (including parts, materials, and consumables related to plant maintenance) will be cycle counted if the item has a part number and a regular on-hand inventory.
- Inventory is divided into the following categories: A, B, C, and critical spares.
- A items are defined in terms of volume and value (or as the SSC publishes).
- B items are defined in terms of volume and value (or as the SSC publishes).
- C items are defined in terms of volume and value (or as the SSC publishes).
- Critical spares are determined to be vital to plant operations using the Critical Spare Parts Algorithm.
- The entire inventory is counted at a minimum of twice per year. This includes daily counts for some A, B, and C items and once-weekly counts for critical spares.

- The count requirement for A items is once every 20 days (M–F only).
- The count requirement for B items is once every 60 days (M–F only).
- The count requirement for C items is once every 120 days (M–F only).
- The count requirement for critical spare items is once every week.

In this element, the SSC will determine:

- How many spare parts and MRO material items the storeroom currently has
- How those line items are broken out in terms of classification (A, B, C, and critical spares)
- How many days will be used for the calculation (M–F only; it is typical to consider a month as 20 days)
- How many line items, of each classification, are to be counted to meet the requirement
- How the random count will be generated, not repeating any number in any classification until all the items are counted once
- Who will perform the cycle count (it is considered a good practice in poorly staffed storerooms to have a technician perform the cycle count with appropriate work order notification and training)

By way of example, Table 5-4 is what the cycle count criteria might look like for an organization that has 2,700 line items in its storeroom with a spread of 15% A items, 35% B items, 48% C items, and 2% D, or critical spare parts.

TABLE 5-4 Cycle Counting Criteria Example

Classification	**% of Inventory Volume**	**# of Line Items**	**Count Every:**	**# of Days in Count Period**	**# to Count Every Day**
A items	15%	405	Month	20	21
B Items	35%	945	Quarter	60	16
C Items	48%	1,296	Semiannual	120	11
Critical Spares	2%	54	Week	5	11
				Daily Count	59

The result from Table 5-4 indicates that to comply with the SSC directive that each A item is counted monthly, B quarterly, C semiannually, and critical spares weekly, approximately 59 items would be required to be cycle counted daily. That seems very manageable.

The three *companion* elements tied to those in the Intermediate-Effective zone are meant to make the storeroom more efficient. That is why you see them in the efficient area of the map. One practice that can drive a highly efficient storeroom is the Optimizing Stock Levels element.

OPTIMIZING STOCK LEVELS

I conducted a very unscientific study and determined that 92% of readers purchase a storeroom-related book simply to figure out how to determine what to stock and what quantity to stock. If you're being honest, that is most likely why you are reading this sentence right now.

The Optimizing Stock Levels element is a strategic practice and is in the efficient zone of the Storeroom Roadmap simply because this is exactly where it belongs.

If you've made the journey with me using this roadmap, you've had years and years of storeroom activity by the time you look to get your inventory makeup and count to match what is truly going on in your facility. You're likely sitting on too much inventory, 50% of which has not moved for over three years, another 20% is deteriorating before your eyes on the shelves, and your management is after you to reduce the inventory value.

You've no doubt begged maintenance to come and help you decide what to get rid of and what you can thin out. You wouldn't have this issue with maintenance if your Stores Stock Committee were engaged and functioning as it should.

Nevertheless, here you are. This is most likely the most consequential element on the entire roadmap, and it is almost exactly in the center.

I think John Moubray's sobering remarks in his industry-standard book, *RCM II: Reliability Centered Maintenance*, summarizes optimizing stock levels and a whole host of other MRO-related thoughts:

> It is possible to use a derivative of the RCM process to optimize spares, stocks and the associated failure management policies. This derivative is because the only reason for keeping a stock of spare parts is to avoid or reduce the consequences of failure.
>
> The relationship between spares and failure consequences hinges on the time it takes to procure spares from suppliers. If

> it could be done instantly, there would be no need to stock any spares at all. But in the real world, procuring spares takes time. This is known as the lead time, which ranges from minutes to several months or years. If the spare is not a stock item, the lead time often dictates how long it takes to repair the failure and the severity of its consequences. On the other hand, holding spares in stock also costs money, so a balance needs to be struck, on a case-by-case basis, between the cost of holding a spare in stock and the total cost of not holding it. (p. 32)

Optimizing the stock levels requires a hard look at the minimum and the maximum of your allowable inventory, the reorder points, safety stock, and other tools that help to control the stock levels.

"Optimizing stock levels" refers to the routine assessment of on-hand inventory and the comparison of that assessment with the recorded usage of those items. Changes in operating demands and the dynamic nature of plant maintenance lead to changes in MRO demands. This process allows decisions and adjustments to be made to the level of stock to be maintained and the manner in which stores will maintain that stock (e.g., min/max, economic order quantity).

The Stores Stock Committee has the responsibility to commission the optimizing stock levels effort. This is the ultimate strategic practice. Nothing in the storeroom sets up your organization for success in the future like getting your inventory strategy correct.

TBR Strategies, the Raleigh, North Carolina–based consulting firm, shows what I believe is the best big-picture view of exactly what we are trying to accomplish. TBR's *Maintenance, Repair & Operations (MRO)* course includes the Order Point Model shown in Figure 5-4 (Day 2, p. 32).

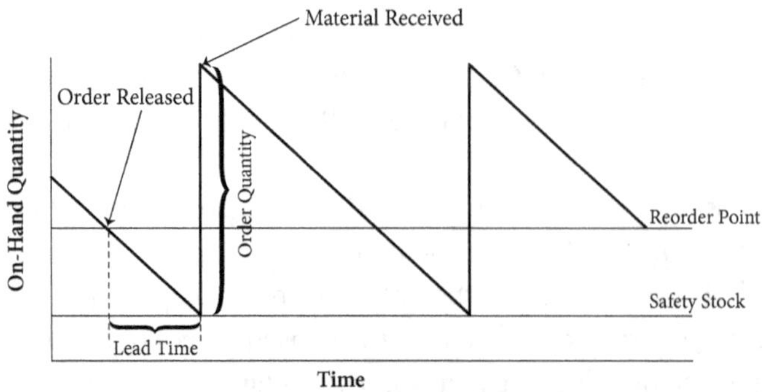

FIGURE 5-4 Order Point Model.

Figure 5-4 has some significant details, most notably:

- **Order released.** In this case, both the reorder point and the minimum desired inventory
- **Materials received.** The time the item ordered is received at the facility
- **Lead time.** The elapsed time between "order released" and "materials received"
- **Safety stock.** The additional stock that is held in reserve, primarily to account for the fluctuation in lead time

Two of the details listed above require further examination.

Lead time is the total time from purchase request to the item received and put on the shelf or into the job kit. It is total time. Very few organizations realize this nuance. If the purchasing department requires many days to release a purchase order, that will count against the lead time. It isn't just the supplier who is at fault for extended lead times. You might have a very inefficient purchasing department. Consider those companies that require their maintenance department to get three quotes before even routine items can be ordered. That doesn't help the lead time.

In a section titled "Hidden Stores (Games We Play by Maintenance and Purchasing)," Doc Palmer suggests that sometimes a delayed purchase order is part of a bigger scheme:

> Some companies with overly simplistic objectives to reduce stock levels might delay approvals or purchase orders. Their reasoning is that delaying the restocking process lowers stock levels overall. That is not necessarily true. In the case of a normal lead time of 1 month to process a purchase order and receive stock, a delay where approvals wait on the manager's desk might extend the lead time to 3 months. In the examples in the preceding section, the increase in lead time from 1 to 3 months should increase the ROP [reorder point] or Min from QOH [quantity on hand] of zero filters to two filters. So, the delay actually increases company costs by increasing the average QOH. The increase in QOH would be the natural progression of discussions between maintenance and stores about how many filters should be kept to promote maintenance efficiency. (p. 328)

Similarly, an inefficient receiving process can drag out a lead time. This was covered extensively in the very first zone, the Foundation-Effective zone.

Last thought on lead time: A consistent lead time is ideal. When materials show up "early," that can result in too much on-hand inventory. When materials show up "late," there is a risk of stockouts. Neither of those two instances is desirable.

The other detail that needs further discussion is *safety stock*. Safety stock is not necessary for all items. In fact, imagine the inventory cost if every item had a safety stock level. Imagine the size of your storeroom! Make no mistake, your company has plenty of safety stock. Right now it is rattling around at the bottom of someone's toolbox as uncontrolled inventory, but it is a type of safety stock. Ron Moore reminds us of the origins of this uncontrolled safety stock as an unintended consequence of unregulated inventory cuts in his manufacturing modeling book, *Making Common Sense Common Practice*. Moore writes:

> If they simply reduce inventory across the board, they could jeopardize their ability to quickly repair failed equipment in a timely manner due to a lack of parts that resulted from reducing inventory, risking unplanned downtime, or incurring extra costs. Further, inventory often will be "reduced" only to find its way into a desk drawer, filing cabinet, storage closet, etc., for future needs. (p. 170)

Safety stock is intended to cover the fluctuation in lead times. Let's face it—sometimes you need to have a little bit extra to make sure you can account for an interruption in the supply chain. Don't increase min/max to cover for this. Instead, understand and utilize the safety stock principles.

Optimizing Stock Levels involves some key concepts. The most important among them are:

- Any item being considered for "optimizing" the stock level and method for stocking should have a history of issuing.
- Storeroom stakeholders will work in good faith to determine the optimum stock level.
- Maintenance leadership and maintenance planners will work diligently to plan and schedule work.
- The intent of this process is to maintain the minimum level of stock necessary to support the need.

- The predictable and stable use of items is greatly desired, as is the use of the Economic Order Quantity (EOQ) model (to be discussed later).

"Minimum" and "maximum" are terms that have been used for decades to refer, respectively, to the least amount of inventory we'd feel comfortable with and the maximum we want. We don't truly want the least amount of inventory because that would be zero, and I don't believe anyone would be comfortable with zero inventory. If we operate with min/max, we are most likely hovering around the max. What that looks like in the many storeroom assessments I accomplish each year is inventory stacked on top of cabinets in the storeroom. The top of a cabinet is not an approved storage location.

The inventory management systems in use today as part of the computerized maintenance management systems have fields in the Item Master Data tables to record *reorder point* and *reorder quantity*.

Not to split hairs, but "reorder point" and "minimum" often mean the same thing. In Figure 5-4, the point where the order was released was also the reorder point, which is the minimum desired stock. But . . . wait. Don't make a mistake in thinking that an order is released exactly at the same time the computer indicates that the inventory has reached the minimum or reorder point. No, no, no. As mentioned above, it may take days for your purchasing department to make this happen. I strongly encourage you to research to determine exactly how long this takes. If you are having lead time concerns, it might stem from the beginning of the process, on your side of the equation.

One storeroom question I get more than any other is how to determine minimum and maximum. That is a fair question and one that I had a concern about when I was in industry.

Before I show some very simple and workable formulas, I'd like to state that, regardless of which agency has ultimate responsibility for the storeroom, the maintenance department is still responsible for telling the storeroom:

- What to stock
- In what quantity to stock it
- What to get rid of
- And what parts are "repairable"

Consider the simple formula for the reorder point shown in Figure 5-5. For simple use, you may also consider this for a minimum level of inventory if you are more comfortable with that terminology.

$$\text{ROP} = \text{d} \times \text{L}$$

$$\text{where d} = \frac{\text{demand per year}}{\text{\# of working days per year}}$$

$$\text{L} = \text{lead time in days}$$

FIGURE 5-5 Reorder Point (ROP) formula.

Use the formula in Figure 5-5 to determine the reorder point for the following scenario.

The Atlas Company runs production 250 days per year. During the year, it uses 60 of a particular seal, which has a lead time of 8 working days. What should the reorder point be for this seal?

You should have calculated a usage rate of 0.24/day. If you multiply 0.24 × 8 days, you arrive at 1.92. Round up to 2 units as the reorder point. Did you get that number? Look back at Figure 5-4, the Order Point Model. On this graph, the reorder point is 2, and the lead time bracket at the bottom is 8 days.

What if we have safety stock? Consider Figure 5-6. Calculate the reorder point if the Atlas Company has a seal in reserve as safety stock. I calculated a reorder point of 3.

$$\text{ROP} = [\text{d} \times \text{L}] + \text{safety stock}$$

$$\text{where d} = \frac{\text{demand per year}}{\text{\# of working days per year}}$$

$$\text{L} = \text{lead time in days}$$

FIGURE 5-6 Reorder Point formula with safety stock.

The maximum is just a slightly harder concept to grasp. That might seem a bit dramatic, but it isn't as simple as you might think. First, consider what having the *maximum* means.

You don't want to exceed the maximum stock level for many reasons.

You could be at the highest level of allowable inventory dollars you can stock. You may physically be out of room to stock any more. You might be dealing with an item that doesn't move or issue out very often, and your concern may be that a lot of inventory will deteriorate on your shelves as time goes by. Much more to the point, is the maximum (in min/max) the same as the reorder quantity?

I look at Figure 5-4 and say yes. The maximum is the reorder quantity. I will admit that not everyone sees it that way. I want to use the last item (not the safety stock) and receive the new order instantly. In that case, I want the maximum and the reorder quantity to be the same.

Figure 5-7 is a simple formula that can be used to calculate the maximum inventory but, as explained above, may also be considered the reorder quantity.

- Min = (lead time [days] × consumption/day) + safety stock
- Max = min + (min/2)
- Example:
  - Lead time is 3 days
  - Daily consumption is 0.5/day
  - Safety stock is 1
  - Min = (3 days × 0.5/day) + 1 = 2.5; round up to 3
  - Max = 3 + (3/2) = 4.5; round up to 5

FIGURE 5-7 Maximum inventory calculation.

Once again, consider the Atlas Machine company and its seals. What would be the maximum inventory (reorder quantity) with and without safety stock? Use the formulas in Figures 5-5 and 5-6. I calculated a maximum of 3 without safety stock and a maximum of 5 with safety stock. (I rounded up from 4.5.) Did you get the same?

Figures 5-8 and 5-9 show a graphical representation of the different min/max results. In the first example, the min/max (reorder point/reorder quantity) is 2/3; in the other, it is 3/5.

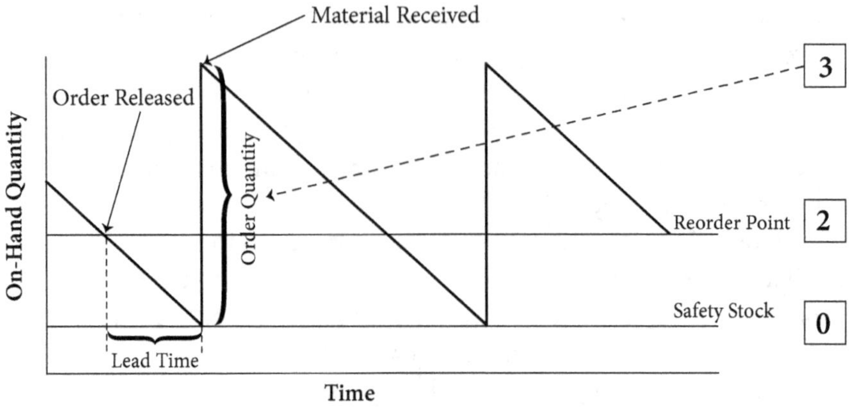

With a usage of 0.24 per day and a lead time of eight days, you would have to order this part for replenishment every four days.

FIGURE 5-8 Graphical min/max image.

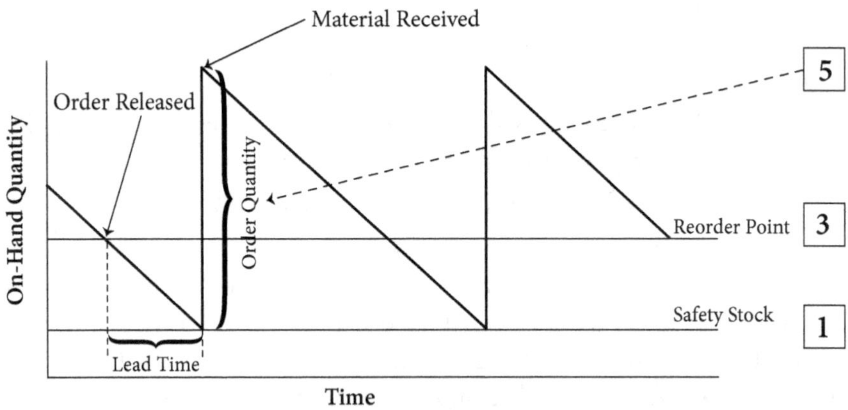

With a usage of 0.24 per day and a lead time of eight days, you would have to order this part for replenishment every eight days. And you would have asafety factor for any lead time disruptions.

FIGURE 5-9 Graphical min/max image with safety stock.

I contend that these are two separate stocking schemes. You and I know that seals don't take up much space or cost much, but there is a difference between the space and cost required to stock three items vs. five items. Also, and this is very much a nuance, the safety stock items must be rotated into usage from time to time to keep them from deteriorating on the shelf. This is a principal use of the *first-in–first-out* (FIFO) methodology.

Thc formula shown in Figure 5-7 is simple and easy to use. It is also not the best way to determine the reorder quantity. I hope I demonstrated some apprehension about making "maximum" and "reorder quantity" synonyms because they are not in all cases. Ideally, the best means to calculate the quantity to order is the EOQ formula.

The EOQ formula shown in Figure 5-10 is from Gulati's *Maintenance and Reliability Best Practices* (p. 140).

$$\sqrt{\frac{2DS}{H}}$$

where 2 = a constant
D = annual usage
S = ordering costs
H = carring costs of an item (carrying cost × item cost)

FIGURE 5-10 Economic Order Quantity formula.

Most people have had experience shopping in the big box stores. One of the selling points about buying products in bulk is that there are savings to be had in buying in large quantities and reducing the number of times you make a trip to purchase an item.

Setting aside any financial discounts for buying more of a product, we must accept the fact that making a purchase costs more because of the cost of the purchasing process itself, which has been discussed in this book. It costs money to generate a purchase order. The fewer purchase orders we generate, the less money we spend on the activity of purchasing. Ordering costs are reduced if I buy more. One last analogy. The fewer runs to the store I make to buy cereal, the more money I save on gas, car maintenance, and time.

However, an offsetting cost must be added to the equation. The more I buy at one time, the more there is to store. I need a bigger pantry to store all the cereal I bought at the box store. Carrying costs rise in relation to the amount of my inventory.

Carrying cost is a hidden cost that will be mentioned many times in this chapter. There is a cost for not having a part and a cost for having a part. Unfortunately, we tend to fret about one, but not the other. In his article, "Use Best Practices for MRO Inventory," Dr. Blache says, "In addition to face-value parts costs, you need to add 20% to 25% carrying costs for factors such as storage space, taxes, damage, and obsolescence. That means, if you keep a part that you don't need, you pay enough to have purchased a new part every four to five years" (p. 6).

These are the countervalues illustrated in Figure 5-11, a graphical representation of EOQ. Notice the brackets around EOQ, positioning at the "sweet spot."

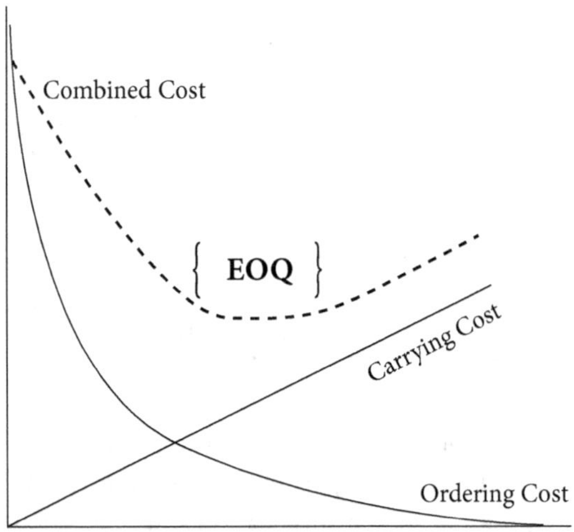

FIGURE 5-11 Graphic depiction of EOQ.

In Figure 5-11, by combining the curve of the ordering cost with the curve of the carrying cost, you arrive at the dashed curve, which is the combined cost. Combining functions graphically was covered in Algebra II in my high school. I confess that it has been a few years. Don't let that deter you from understanding that knowing the EOQ formula is valuable. It is important to know that EOQ calculations are usually only meant for items that have a stable level of use. Erratic or highly fluctuating demand for some items makes them poor candidates for this ordering method.

Let's revisit the Atlas Company and its 60 seals. Using the numbers that we know, add the following factors:

- The ordering cost is $65 a purchase order.
- The seal cost is $8.
- Carrying costs are 25% (0.25).

Figure 5-12 is my attempt at using the EOQ formula.

2 = a constant
D = 60
S = \$65
H = 25% × \$8 (0.25 × \$8 = \$2)

$$\sqrt{\frac{2(60)(\$65)}{\$2}}$$

FIGURE 5-12 EOQ formula example.

What did you get for an answer? I got 62.4. According to the EOQ formula, I would be wise to buy all 60 of the seals at one time. This makes plenty of sense if I were to use all 60 at one time, but in my example, that wasn't my intent. I intended to set up an example showing that 60 seals are used yearly. A little here, a little there. How did the EOQ arrive at 60 being my best bet?

Go back to Figure 5-8. With a maximum of three and a minimum (reorder point) of two, I would have about four days of inventory before I had to reorder three seals (daily usage of 0.24 × 4 days = 1 item, roughly). So according to Figure 5-8, every four days I would order three seals. That's three seals at a cost of \$24 (\$8 × 3), and I would pay \$65 to cut the purchase order. Remember, it cost me \$65 for every purchase order. I would do this 62 times a year (250 days divided by four days of inventory). Using Figure 5-8, I could spend over \$4,000 on ordering costs for \$480 worth of seals in a year. It doesn't get too much better with Figure 5-9.

If you had to go to the grocery store every four days to buy cereal, wouldn't you buy more than one box of cereal?

Economic Order Quantity isn't for everything and everybody. It is part of the Optimizing Stock Levels for reasons and useful for those circumstances that make fiscal sense.

Another optimization methodology used quite often is Order on Request (OOR), which has been discussed to some extent in the earlier chapters. It is very likely that an organization has more stock numbers for items stocked as 0/0 (0 min/0 max) than actual items stocked on the shelf. Take advantage of this opportunity.

Due to low turn rates and the ease at which some items can be attained, it might make more sense to set the items up in the Item Master Data and assign them a stock number. The planner can order the items by part number. When they arrive, they will be received directly into the job kit following a process that has been outlined many times already.

This element is meant to drive efficiencies in the storeroom by evaluating how goods have moved through the storeroom. After your storeroom has been in service for a period of time, the Stores Stock Committee will determine that it's time to evaluate if the stocking levels are correct and will specify the best methods for optimizing the stock levels. Because this is a strategic practice, performing this evaluation more than four times a year is not recommended. I wouldn't recommend it more than once per year.

Phillip Slater offers a word of caution to those thinking that optimization will solve all your inventory problems. In a comment about understanding how to reduce inventory levels, Slater writes, "Only the ICR [inventory cash release] approach will help you identify how to reduce your inventory levels by adjusting your management techniques and processes" (p. 169). Slater goes on to explain that ICR is used to "determine the most appropriate initial order quantity" (p.106).

There are two additional organizational elements that we've hinted at their use:

- Requisition of Materials
- Receiving Purchased Items from Stores

REQUISITION OF MATERIALS

This organizational element is half of a tandem of elements included in a non-stock items category. I cannot stress how important it is for your organization to have clear guidelines on who can purchase items received in the storeroom. This should be elementary, but it can be very confusing and very costly.

Imagine a company where several people have purchasing authorization. A moderately sized facility might have close to 100 people who can purchase items. All these purchases should be received through some receiving process. Not all of them come in through the storeroom, but they should at least come in through central receiving. If this is the practice, and there are no guidelines or restrictions, then almost anything can show up.

Several years ago, I was consulting in an Idaho-based company, and while I was talking with a storeroom associate, he opened a box and pulled out a full-length mink coat. This had been received into the storeroom. The person who ordered the coat was not at the plant at the time, and the coat had to be stored somewhere until the matter could be sorted out. Apparently, the employee had ordered the coat as a present for his wife but didn't want it delivered to his home to avoid ruining the surprise.

I use this absurdity to illustrate an absurdity. The storeroom should only be receiving storeroom-related items. Typically, that would be spare parts and repair materials. There should be a well-established and greatly communicated process for authorized personnel to follow to requisition materials that will be received in the storeroom. Once a part received in the storeroom is reconciled with the purchase order, the part goes on the shelf as replenishment or in a job kit. Where does a mink coat go?

For non-stock spare parts or materials, the planner will identify the part or material that is needed for a work assignment. The planner does this by listing the item details on the work order. Stocked items are referred to by part number, and 0/0 items are referred to by part number as well. Still other items might not be stocked or have stock numbers. These are truly the items that you are purchasing on demand. For these items, the planner will include all the information that will be needed to issue a purchase order. The planner will most likely be purchasing the part from an approved vendor.

Take time to learn and understand how your inventory and work management systems work together to complete this greatly exercised process. This should be a very smooth part of your computerized maintenance management system.

RECEIVING PURCHASED ITEMS FROM STORES

This may be the shortest section of this book. This process is the second half of the tandem of organizational elements that make up the non-stock items category. Recall from the Receiving element that items received into the storeroom either go onto the shelves as replenishment inventory or go into a job kit. That is still the case.

If you further consider non-stock items being received, it is understandably important that the purchase order generated for the item has the work order number. If the work order number is not on the purchase order, it will not be clear what job the item is for, and the item could easily be confused as needing to be stocked (even though it is not set up as a stocked item).

I'm trying really hard not to say that the storeroom gets a lot of junk delivered, and that the associates spend a lot of time figuring out where it goes. If you order something that gets received through the storeroom, play by the rules (see "mink coat"). It's a privilege, not a right. Second, for the storeroom, the rules that are established should be rules that people will follow.

As you'll notice on the Storeroom Roadmap, some elements were tied to the Intermediate-Effective zone in this zone for the Intermediate-Efficient. Then there are others, a majority, that seem to blossom from the Foundation-Efficient zone. We just finished dissecting the former; let's look at the latter.

PARTS STANDARDIZATION

Because you are well into this book, I can confess to you that there are two secret means to improve the efficiency of your storeroom. Make no mistake, you need all 32 processes, practices, KPIs, and organizational elements. But to increase your efficiency, it comes down to optimizing your stock levels and standardizing your parts.

Figure 5-13 outlines the workflow for a very successful Parts Standardization process.

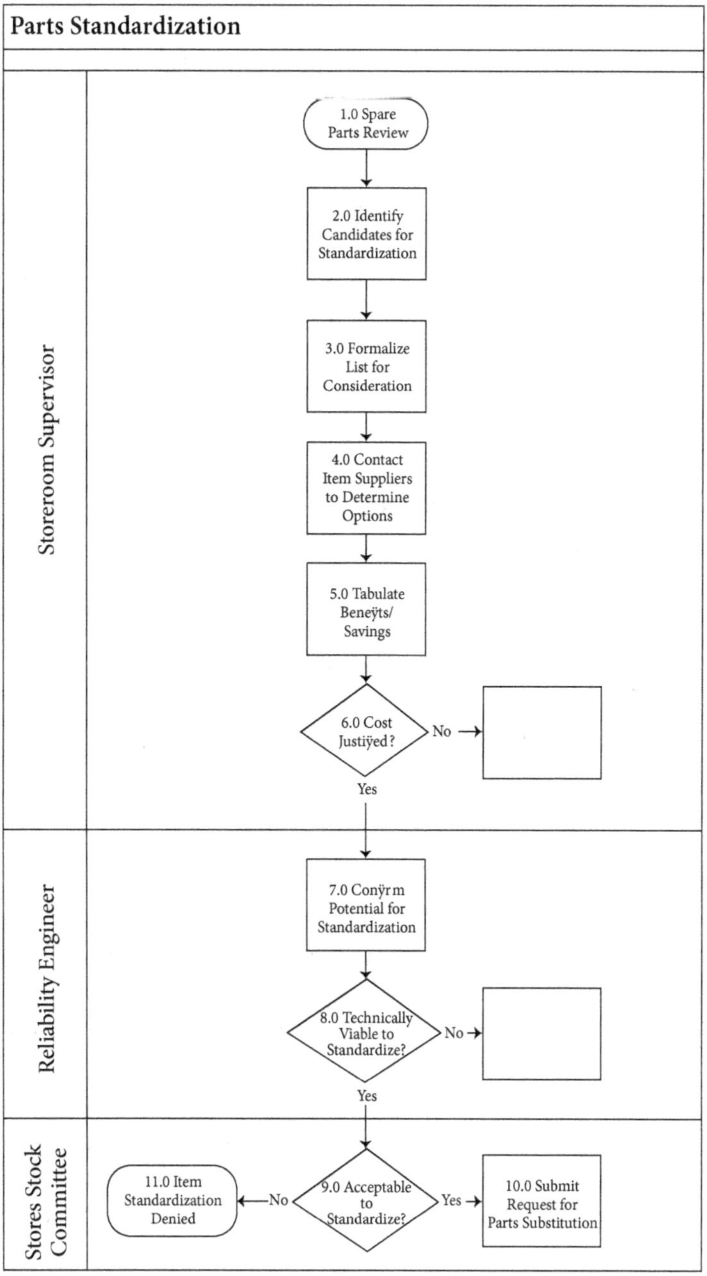

FIGURE 5-13 Parts Standardization (partial).

Once you understand the power of this process, you will be hard-pressed not to execute it. It makes perfect sense to dovetail this element with the foundational element of Data Scrubbing.

One of the benefits of *scrubbing* your *data* is that you can confirm that all the information for each item is complete in the Item Master Data. The data tables in most CMMSs can have up to 47 fields to record detailed information. During the analysis of this information, storeroom personnel will notice that they have duplication of parts. Some estimates are as high as 11% duplication. How does this happen?

The Parts Standardization process requires effort by the storeroom associates to keep inventory value and line-item totals to a manageable level. For parts standardization, maintenance, procurement, and the storeroom must work together to see the commonality in many "like items" and take steps to do what is logical and necessary to pare down the inventory to one type of item for combined purposes.

Take the scenario where a manufacturing plant might stock Timken bearings because that is what its supplier carries. Without a thorough New Item Setup process, a mechanic can request that SKF bearings be added to the inventory. Another mechanic might suggest stocking NSK bearings. Before they know it, they will be stocking three "flavors" of the same exact bearing.

This is real, and it happens quite a bit. I suspect those of you reading this passage are aware that you have many duplicates, and even more that you believe are duplicates. This costs a tremendous amount of operating capital (tied up in spare parts) and increases the shelf space needed for storerooms, not to mention all the extra items that must be cycle counted and cared for. There is no good that can come from nonstandardization.

Aside from poor stock addition processes, how do companies end up with nonstandardized parts? It starts where a lot of ills begin . . . with engineering.

Your organization should standardize around the make, brand, and model (if possible) of components that seem to work around your crazy place. Design and project engineers should have a formal template declaring (as an example):

- We are an Allen-Bradley company.
- We use SKF bearings.
- We use Dodge gearboxes.
- Etc.

I get it. Projects are completed, new capital assets are designed on a budget, and you get the parts that your OEM gets the best deal on. The maintenance department and your organization will be the victims of this short-sightedness for decades to come.

Electronic components (4–20 mA) fail *randomly* and *suddenly*. You cannot perform any meaningful preventive maintenance on an electronic part. The failure is going to be on a tiny silicon chip, on a circuit board, in a slot, with a panel door shut, in an electrical panel with that door shut. By the time you get a whiff of ozone or a metallic taste in your mouth, the part is already fried.

Your only strategy against being dead in the water for electronic failures is to have a spare in inventory. Since your company didn't promote standardization, you'll need to stock three types of PLCs (programmable logic controllers) and seven types of VFDs (variable frequency drives). Since these are critical electronic components, you will want to store them in an air-conditioned building.

You get the picture.

Whether standardizing at the front end with effective and good engineering practices or using the results of data scrubbing to weed out duplicates, there is tremendous value in reducing stock count.

The standardization mandate is a core discipline in total productive maintenance (TPM). Figure 5-14 is a picture of the TPM puzzle shown in Marshall Institute's *TPM/TPR Coordinators Workshop* (p. 33).

- **Shop Floor Involvement**
- **Standardized Parts**
- **Documentation**
- **PM, Parts, and Training**

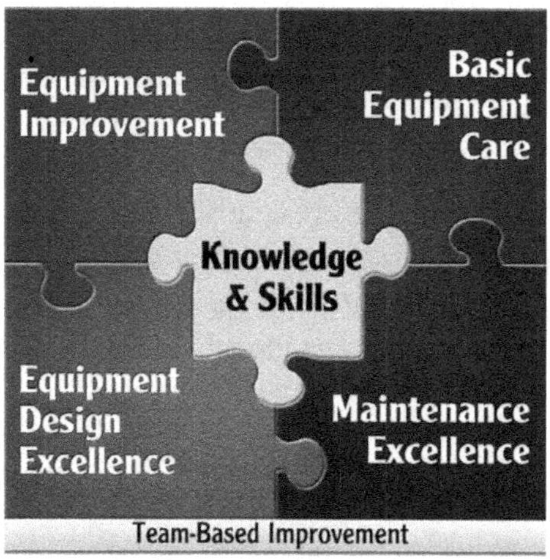

FIGURE 5-14 TPM puzzle.

I strongly encourage every organization to analyze asset performance and component life-spans. Determine what is needed to get at least the estimated life from your components, and focus on the parts that seem to work for you. Understandably, if your organization has grown through acquisition, standardize on parts that work for the different divisions. Don't standardize for cost savings. Just don't do it. Performance . . . only. The cost savings will come from standardization.

Doc Palmer brings the value of standardization home in this passage:

> A management convinced of the overall company benefit should allow maintenance to dictate certain equipment standards for purchasing and engineering to follow. Maintenance management should take on such a course of action as an important responsibility, not one simply to be recognized but to be continually managed.
>
> Many plants have decided that standardization is the most important key to reliability, high capacity, and profits. These companies have standardized and reduced entire assembly lines or processes to standardize as much as possible. They have even standardized or restricted allowable suppliers and vendors to a select few with whom they develop close relations and high expectations of consistency. (pp. 340–341)

The last sentence in Doc's message—reducing suppliers and expecting more from them—is right out of W. Edwards Deming's playbook. Even Palmer makes this observation, writing, "[Deming] for quality reasons alone recommends having long-term relationships with single suppliers . . ." (p. 342).

In describing an example in his modeling scenario, Ron Moore points out that "suppliers were consolidated and standardized as much as possible" (p.162). He makes that observation as a positive attribute.

One other element is heavily leaned on during the standardization process, and it's not Data Scrubbing, although that is as well. It is the organizational element of a Bill of Materials.

BOM

I wrote earlier that the Storeroom Roadmap hadn't changed much since its inception, but that is not true. During the years since Stephen Miller and I first discussed this idea, and I first documented the layout of the roadmap, I have made some observations and come to some conclusions. Those altered the map a bit. Keep in mind, and I wrote this earlier, the map is my opinion. I've done a fair bit of research on storerooms, and I think it is a very learned opinion.

One very noticeable change concerns this element. Originally, the Bill of Materials element was a tactical process. This would be a monthly concern for the two parties, typically maintenance and the storeroom. I regard this element as more of an organizational need and less of a process. If there were a process aspect, it would be that the maintenance planner is the sole person responsible for keeping the BOM updated. As an organizational element, it speaks to a much broader need and to a much broader audience.

Not only is the Bill of Materials an organizational element with the potential to impact every agency in your organization (see Figure 1-6), but it is also a steadfast rule (see Rule #2).

Much has been written already about the BOM, and I don't believe I can add much more to the knowledge we've already shared. However, there are some points to make as you consider the practical location of this element on the map.

The triad formed by the connection of Data Scrubbing–Parts Standardization–BOM should make it overtly clear to you that there must be agreement among the three.

The storeroom is not going to stock any item that doesn't appear on an active asset's bill of materials. When we are standardizing parts and materials, we should look at what is on the BOM for each asset. Any change from what is on an asset's BOM to what you will be stocking in support of that asset must match. That didn't come across as clearly as I had hoped. Let me try again. If you make an engineered change to an asset and start using a different component, that change will be reflected in the BOM for that asset as well as the inventory management system. The new component in the inventory management system must reflect that it "goes to" this asset.

If you haven't already made these types of changes using an MOC (Management of Change) process, I suggest you start. Immediately. All the part or material information will be spelled out in detail in the Item Master Data, which is the subject of the Data Scrubbing element.

Look back on all that you have done on your journey so far. You have established some solid processes and practices during your trip. You have recently honed your inventory, sharpened your efficiencies, and even whittled down your supplies through standardization and a look at min/max. With a decrease in what you are carrying and optimization of what you have left, it might be time to look again at how your storeroom is physically set up to function.

OPTIMIZED PHYSICAL LAYOUT

More than any other single element, this is likely where I'm called in to engage with storerooms at the beginning of any relationship. Common areas of interest from prospective clients are their desire to:

- 5S the storeroom—sort, set in order, shine, standardize, and sustain
- Add vending machines to their program
- Switch from a dedicated room to vertical lift modules (VLMs)
- Relamp the storeroom
- Set up satellite locations
- Convert to using bar coding
- Etc.

Imagine if you had started the storeroom journey at this node on the map. How much improvement would have been made so far? I suspect none. Better yet, how sustainable would anything you put into place honestly be? If you guessed "Not very sustainable," you'd be right. I suspect you have tried this yourself and had similar results.

These are excellent ideas and great practices, but they come when a storeroom has matured to the stage where these are the next steps to take in the logical evolution of a storeroom—a storeroom that has established a discipline toward purposeful work execution and is working as a team with the stakeholders. You don't simply parachute into this element. I tell my classes and my clients that if you can't master pencil and paper, technology isn't going to work for you.

On the assumption that you've reached this element in the Intermediate-Efficient zone by way of the established paths discussed so far in this book, it is time to look at your storeroom in terms of its physical layout and the ability to provide amplified service and convenience.

For example, at this juncture, your organization should be highly engaged in kitting for planned maintenance jobs. We will get to a small process in the advanced zones called Delivery. This is the delivery of kitted items. If you've been working as a team to solidify the planned and scheduled job functions of the maintenance department, and your storeroom has had focused success with job kitting, it might be time for your group to look at delivering the kits to the jobsite just before the time they are required for use. That future service might require a slightly *different-looking* storeroom. More on the Delivery process later.

Some major areas to concentrate on and develop in the Optimize Physical Layout organizational element include:

- Physical location of the storeroom
- Point-of-use devices and methodologies
- Lubrication storage
- Sensitive-component storage
- Non-materials storage

When I attended my very first MRO storeroom training course, the instructor had a slide in his PowerPoint presentation indicating that the storeroom should be located near the maintenance shop. I remember questioning this because it seemed to me that the parts in the storeroom would more likely be used in the production areas and not in the maintenance shop. "Wouldn't it make more sense to put the storeroom in the middle of production?" I asked. I was shot down pretty quickly.

In my very first civilian job, I was told by my boss to find some space to build a storeroom, as we didn't have one, and I had lobbied for funding and resources. "Pick any place that makes sense," my boss told me. I chose prime production real estate and got it.

I don't know if anyone else has ever lobbied for and gotten square footage near production, but I can't help but think that makes a lot of sense. The idea is to increase efficiency. If I put the product near where the product is used, I cut down on one of the seven forms of waste. Travel.

When the time is right for your storeroom and the Stores Stock Committee to determine what ways the storeroom can improve its service, the physical location might be the topic to discuss.

In lieu of moving the storeroom to a better location, it might be more practical to consider the different types of point-of-use methodologies that are commonly brought into service. Many of these are yesterday's ideas but keep delivering today's value:

- Satellite storerooms
- Vending machines
- VLMs

As the name implies, the *point of use* intends to put the items as close to the point of use as possible. Care must be taken to ensure that the inherent reliability of the part itself is not compromised by the means of storage and distribution, and to guard against loss, shrinkage, and just outright theft.

I'm sure satellite storerooms have been used since the Industrial Revolution. What started off as likely metal (or wooden) shelves in the corner of some production cell evolved into cabinets and even small out-buildings inside the facility building to house parts and materials that were likely to be used locally. These were almost always unstaffed and would often be unserviced. Parts for satellite storerooms would typically come from the main warehouse or storeroom.

Although the use was often undisciplined, and the satellite locations were abused and unkempt, their need was and is today still quite necessary. Satellite storerooms are most needed in facilities with tremendously large operational areas. This is common in oil and gas production and processing facilities, chemical plants, universities, and other larger and complex facilities.

Satellite storeroom locations must be managed, and the items must be kept in good order. This requires great teamwork and responsibility for the storeroom associates and those who receive value from the locations. Remember that the inventory in these locations is still protected inventory and should be treated as such.

I can recall seeing my first storeroom vending machine. It was a converted candy machine with only first aid supplies. I believe, and I'm trying to recall exactly, but I think this machine also had chewing gum. That's right. Next to the aspirin, bandages, and gauze patches was spearmint gum. I honestly didn't see the value in this.

For decades, it seemed, storerooms were putting vending machines out in the production facility and only stocking these machines with non-MRO materials. It was not unusual to see safety gloves, hair nets, first aid items, and maybe some small drill bits or end mills in a vending machine. Where were the actual spare parts? Some companies even put the vending machines immediately outside the actual storeroom. This was a location that I thought negated the point-of-use value.

The last place I ever worked as an employee was in a food processing plant. I was the plant engineer, and the storeroom was under my jurisdiction. We couldn't put vending machines in the food processing areas for sanitation reasons and for food safety reasons, but we had exactly one machine, and it was near the employee break area. That machine contained Kevlar gloves. Management had discovered that some employees were taking up to five sets of Kevlar gloves from inventory every single day. This was costing the company thousands of dollars a week. The vending machine was installed for the sole purpose of tracking usage. This is not the primary purpose of a vending machine, but it is a good secondary reason.

Again, the idea's main focus is on the point of use. What are some MRO spare parts and materials that make sense to locate close to the locations where they would be used? Many companies, and probably yours, have a hard time determining what to put in a vending machine. One of your storeroom suppliers has likely been willing to help you make that determination. I'm not so sure that is the best approach. I can imagine that if you have storeroom vending machines, some of them are sparsely stocked. I could be wrong.

The idea and philosophy of point of use should be arrived at with some research and lots of logic. Many organizations rush into the idea of vending machines because "everyone else has them." Might I suggest that this is not the best rationale?

The SSC is the working group that should be called upon to determine if vending machines are right for your operation and if the time to engage is now. Perhaps the best execution of the point-of-use ideology and practice is at a manufacturing facility in the upper Midwest. This company has a storeroom but no attendants. Its operation is so advanced that the company doesn't need them. The company also has the most elaborate series of vending machines throughout its facility that I've seen. These machines have actual MRO spare parts and materials, cordless power tools, and other support items. Figure 5-15 is a picture of one of its vending machine complexes.

FIGURE 5-15 Vending machine complex.

Vending machines are common and have great value if used correctly and for the right reasons. Another common storage device that mature storerooms evolve into is storage bins. It is dated technology, but bin storage offers great value for compact storage and for the ancillary fact that items are kept clean in a bin.

Some companies have combined the accounting benefit of a vending machine with the compact storage that a bin provides into a hybrid vending bin. Figure 5-16 is a picture of the very first vending bin I ever encountered. This was in use in the upper Northwest at a Fortune 100 company.

Satellite storeroom locations and vending machines are meant for items that turn frequently. There is no value to having slow-moving items in a faraway, publicly accessible location. I don't advise it, and I can't imagine that anyone would stock expensive and critical items in a vending machine or in a faraway satellite location that is loosely guarded. If you employ these types of service locations, you might notice that you have a collection of fast-moving, generally inexpensive items for offer. Those are the types of items typically made available for these kinds of delivery mechanisms.

There is no value to a maintenance technician having to walk back to the storeroom to get something; likewise, there is no value to an operator having to walk to the storeroom to get something. For that reason, satellite stores and vending machines should house maintenance and operating supplies. I strongly encourage you to make sure the SSC is engaged in recommending the kinds of items to make available.

FIGURE 5-16 Vending bin.

I ask a simple question in every one of my storeroom classes: "If your storeroom inventory burned up tonight, just the inventory, not the building, and I gave you a check tomorrow for the full value, would you repurchase every item one-for-one, and would you lay it out in exactly the same way?" No one ever answers yes to either part of that question. This is your chance to stock different *stuff* and lay it out in a completely new way that works better for you and your team.

You aren't being asked to just move the furniture around in this element. The storeroom is being challenged to establish a leaner, service-focused environment that sets the MRO effort up for success well into the advanced zones of the map. You cannot have this discussion today in any meaningful way without having accomplished all the work you've done so far. Just look in the rearview mirror one more time to see the progress that has been made.

It seems that a lot of the technology that has been employed to get parts to the customer is somewhat dated. Yet, many ideas are making a comeback and being looked at once again. A recent client informed me that he is considering getting rid of the storeroom altogether and going exclusively to vertical lift modules. Make that two clients in about a year who have told me this. VLMs seem attractive until they aren't.

I encountered my first VLM in 1989 at RAF Lakenheath, an American Air Force base in England. I was stationed there in the late 1980s. The base had a central warehouse, and one day I had an opportunity to go there for something. In the warehouse, the logistic folks had this very tall machine sitting right in the middle of the building. A machine like no other machine I had ever seen. I asked the sergeant what the machine was, and he told me that it is where they housed all the supplies. It wasn't called a VLM back then, or maybe it was. I just didn't hear that term.

I don't know exactly when this technology was developed. I just know that it's been around since 1989. I saw one. The technology must have been relatively new, as the sergeant told me the machine was an absolute nightmare. On many occasions, they had to use a man-lift to go up the side of the machine, remove panels, and either get parts or realign something so the "thing" would work.

I suspect that a lot has changed over the decades, and I see more and more of these devices in use. I should stress that, like satellite stores, vending machines and other point-of-use applications make sure that the passion and energy are always pointed toward providing the best service and convenience. Right part, right time, right quantity—that is the goal of the storeroom. Regardless of technology.

In the final analysis, the storeroom layout should reflect a well-thought-out approach, adequate space, and well-labeled products. Figure 5-17 is a good example of what a high-performing storeroom might look like.

Sometimes the item needed isn't a part or material per se; it is an oil or other lubricant.

I've mentioned in my other books that the most important activity that maintenance can engage in is preventive maintenance. There are seven types of preventive maintenance. The most important type is lubrication. That makes lubrication the most important activity that maintenance can be involved in.

This point is so critical to the effective delivery of sound maintenance that when I conduct maintenance and reliability assessments and storeroom assessments, I also take a tour of the oil storage area. The way the oil storage areas look and are maintained provides insight into everything you need to know about the function and discipline of the maintenance organization.

FIGURE 5-17 Storeroom shelving.

Organizations are encouraged and most likely required to consolidate oils and lubricants to a significant few. I recommend having no more than three unique oils and greases. Noted reliability consultant and author Preston Ingalls spells out 14 Greasing Tips, or what he refers to as "Making Peace with Grease" in his publication *Making Friction Fiction*. Ingalls offers tip number one: "Minimize the number of greases by consolidating brands and types" (p. 22). His book goes on to discuss the ills of cross contamination and incompatible lubricants. Risk is reduced if there are fewer opportunities to be at risk.

Some companies have assets that require unique oils and greases, and there are exceptions to the consolidation concept. It is easier to manage a few oils and greases. Most have found the practice of consolidation to be quite helpful. I further recommend engaging with a lubrication consulting service to make this happen. There are too many opportunities to get this wrong unless you are an expert.

Although I have never had what I consider an extremely good example of an outstanding oil and lubricant storage area, I have seen one. If you've attended any of my training courses, you have heard me speak about this facility. A black-and-white picture doesn't do this facility justice, but review Figures 5-18 and 5-19, and get a sense of the order and orderliness of this lubrication storage area.

FIGURE 5-18 Lubrication storage.

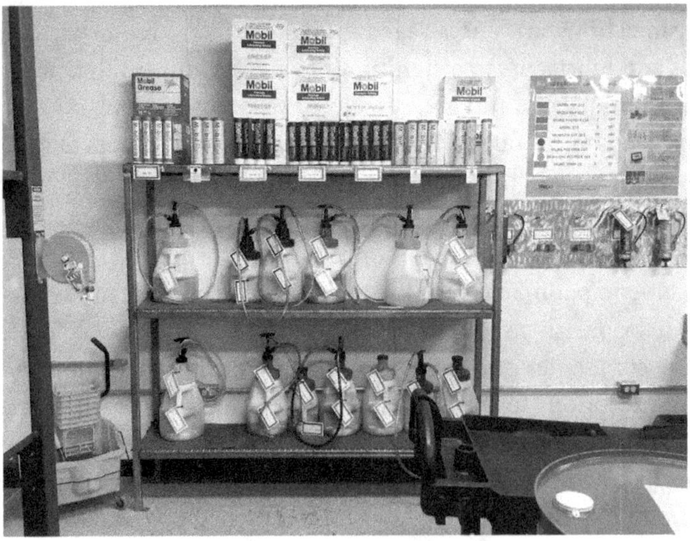

FIGURE 5-19 Lubrication storage.

You don't want a storeroom that looks good and functions horribly. And you don't want a lubrication storage area to look great, but the delivery of service and product to be less than ideal. I can tell you that the lubrication program featured in Figures 5-18 and 5-19 is a global award winner for having the world's best lubrication program (storage and practice). This is truly a remarkable operation.

Like having a dedicated location for lubricants, the same applies to sensitive components, especially electronic ones. Heat and humidity can certainly damage electronic components that are in operation. I don't suppose the presence of heat and humidity aids in maintaining the inherent reliability of components while they are in storage. For that reason, it is recommended that the 4–20 mA components be stored in a dry, cool place.

One of the very first storerooms I ever was part of creating from scratch was on the Gulf Coast in Louisiana. If you've been down to the Gulf in Louisiana in the summer, you have likely witnessed rain dripping from the ceiling due to the humidity hanging in the air. This storeroom was going to be constructed with a building inside the main storeroom. The main storeroom was going to be air-conditioned for the comfort of the employees, but the smaller building had a dedicated air conditioner and dehumidifier to guard the sensitive and precision components. I thought this was an ideal setup. Electronic component OEMs will be able to provide detailed information on the temperature and humidity that is ideal for storing your parts and materials.

This storeroom assignment was also where I sparked my elevated interest in storeroom operations and started the creation of many of the process guides and workflows you've seen in this book. Those workflows and associated process guides have matured and changed over the years to include best practices I have seen and tried. I want to point out that the storeroom manager in Louisiana was a very competent leader, and I learned as much from him as I believe he learned from me. *Side note:* Never stop learning.

When you arrive at this element on the Storeroom Roadmap where you are to evaluate the best layout and operation for your storeroom's efforts of the future, you also need to consider some of the non-material storage elements that are so often part of the storeroom service:

- Special tools
- Safety items
- Technical library

SPECIAL TOOLS

It might be hard to imagine that Special Tools is a strategic practice, but it is one that most storerooms are familiar with. This is simply the practice of storing shared, common tools in the storeroom. Sometimes this location is called a "tool crib." However, for those in the machining industry, you know a tool crib to be something different.

I first encountered the Special Tools element as an officer in the Air Force. I was an aircraft maintenance officer and worked on the flight line. Our crew chiefs didn't have their own tools; they checked out a crew chief toolbox from the storeroom. They would also check out any other shared or common tool or device as needed.

The highly nontechnical way the storeroom folks could keep track of "who had what" was accomplished by using *chits*. I've seen this practice in many locations around the world. If you want a tool or device that is controlled by the storeroom, you have a ring of stamped metal discs, each with a unique number assigned to you. You give the associate in the storeroom one of your chits in exchange for the item. The chit is then placed on a peg where the item was. When you return the item, you get your chit back. Figure 5-20 closely approximates what our chits looked like. Each of us had our own individual number. As hard as this is to believe, in the Air Force of the 1980s, these chits were kept on your belt loop by a shower curtain ring.

FIGURE 5-20 Chits.

In aircraft maintenance, no aircraft is released for flight until all the tools are accounted for. I don't believe I've mentioned this, but I was also an aircraft crash investigator in the military. I have been on a few incidents that were caused by tools jammed in the controls. Some of you might deal with circumstances where the presence of foreign objects can result in products being quarantined and even possible recalls. Accounting for tools is a serious business.

Do you recall the mention of the 5S program earlier? The second S, the one for "set in order," has two meanings. The first meaning clearly indicates that everything has a place, and everything should be in its place. That is likely as far as some people consider this to mean. The second meaning is even more important. The item being stored or put back into its place is required to be serviceable. It is the responsibility of the person returning the item to storage to ensure that the item is ready for use the next time. You work in an industrial setting, and it would be understandable to assume that from time to time, items will be broken while in use. If this is ever the case, record the incident and replace the item.

I would also like you to look back at Figure 5-15. This figure shows a vending machine complex. This photo tells an even bigger story. The storage boxes in the foreground are full of batteries. Further to the left in the picture are the actual cordless tools. This company was having issues with technicians putting tools with dead batteries back in storage locations. They were complying with the first part of the second S, but not the second part . . . ensuring that the item is serviceable. This company added vending slots, drilled holes in the back of the boxes, and put batteries and chargers in each box. The holes were for the power cords. Life will find a way!

In my presentations, I used Figures 5-21 and 5-22 to show a good special tools operation within a highly effective storeroom.

FIGURE 5-21 Special tools.

FIGURE 5-22 Special tools.

This must be a practice where the tools and devices are checked for serviceability before they are issued, and they are checked for cleanliness and serviceability when returned.

If you recall, a strategic practice is one that is exercised no more than four times per year. Surely, your organization checks out the torque wrench more often than that. This element is a strategic practice because, at this position on the map, we are not referring to the actual exchange of part for chit and chit for part, but rather as it relates to the storeroom layout optimization. We are optimizing the tools we store and the technology we use to ensure tools are tracked and confirmed to be serviceable at all times (e.g., vending machines).

This element is the proper time that the storeroom organization must determine what resources are needed to provide even greater reliability-related services. I can tell you that a storeroom that can't keep track of a gear puller will have a hard time keeping up with an optical alignment kit and all its pieces.

Special Tools is the only practice element that makes up the non-materials storage category. There are two organizational elements.

SAFETY ITEMS

Hopefully, you can see the important nature of storing special tools and confirming their serviceability before and after each use. That same approach should easily be applicable to safety items. It should be clearly understood that storing a safety device in the storeroom does not absolve the user of the responsibility to ensure it is in good order, safe, serviceable, and used properly. It is 100% likely that safety devices are stored in the storeroom for convenience and it is the only place that is somewhat under lock and key.

Such safety items might include:

- Fall protection harnesses and lanyards
- Personal protective equipment (PPE)
- First aid supplies and kits
- Spill kit materials
- Locks and tags for LO/TO (lockout/tagout)
- Fire extinguishers
- Confined-space entry equipment (air monitor, tripod, etc.)

Your organization might have a combined HSSE (health, safety, security, and environment) department.

Because safety items are commonly stocked in the storeroom, a safety department representative is requested to be on the Stores Stock Committee. The safety department is, after all, a stakeholder in an effective and efficient storeroom. The safety manager and the manager's team cannot be allowed to simply put stuff in the storeroom without using the proper item-setup protocol. Likewise, any safety item formally established as a checkout item must have instructions of care and issuing of the item. This must come from the safety department. Too often, the safety department has had a hands-off approach to safety items. This indifference should not be tolerated in an organization. The safety department must be engaged in the transactions and in monitoring if the processes for these life-critical items are to be managed and executed precisely right, every time.

How strong is your safety department? Does it operate proactively to ensure that the safety devices stored in the storeroom for common use are safe and serviceable? Does the department work in partnership through audits and assessments to make sure all the devices are usable? Does the department replace worn and damaged items? Is it clear how to proceed when a needed item is unserviceable? This last question is the one that you probably have faced more than once.

If your organization insists on stocking safety-related items in the storeroom, the safety department should be front and center representing the availability and serviceability of those items.

Make a strawman argument against such a policy, and see if it makes sense. I contend that if you mandate an activity that someone else must comply with, you have a responsibility to be engaged in the execution and sustainability.

I haven't even spoken about material data sheets (MDSs). If your organization provides a paper copy of the MDS, a copy should be in the storeroom. Again, your HSSE department has the responsibility to help keep these up to date and not just point out faults and errors when they occur. (See the definition of "proactive" in Chapter 1.)

TECHNICAL LIBRARY

Look back at Figure 2-2, which shows an example of a storeroom layout. In the upper left of the figure are three dots representing three individuals sitting at their desks. These people are the maintenance planners. To the right, in the room where they are sitting, is the technical library.

It isn't necessary, or even recommended, that the maintenance planners relocate to the storeroom, but this client had that belief. What is certain, however, is that the planner has the oversight responsibility for the technical library. Further, the technical library should be in the storeroom.

In this century, a lot of the technical materials and information for your assets (schematics, drawings, piping and instrumentation diagrams [P&IDs], BOMs, etc.) are likely to be electronic. But there is still a lot of technical information on your company's assets that is likely to be in books and on paper drawings. All this information should be brought into one central library. That is the technical library. On a side note, I would recommend digitizing this information.

It would not be uncommon for a lot of your technical information to be outdated. Perhaps some of your drawings have been red lined, but not updated. Maybe your equipment has been modified without noting such modifications on the drawings. There have been a lot of less-than-ideal practices happening over the course of many years. Regardless of the state of your technical manuals and materials, they still represent a wealth of information that can be mined for the benefit of the organization.

The planner, as stated, is responsible for the technical library. The reason to locate it in the storeroom is twofold.

First, the storeroom should be a secure storage location. This is useful for guarding the only sets of technical information that might exist. The storeroom has limited access, by invitation only. The maintenance planner has carte blanche access as the liaison to the storeroom. Never, ever let a technical library document out of the library. Instead, have a very good copier in the library.

Second, the planner can use the space for research when collecting information on parts to order that might not be stocked. It is the single and central reference library for the planner and the storeroom associates to utilize when conferring on a parts and materials plan.

Think about your organization's technical library. Is it set up and guarded like the reference library for a college or university? In most colleges and universities, you can't check out reference materials. They are required to be reviewed in a secure room. You aren't even allowed to take the material out of the room. That is how it should be.

Before we leave the technical library, astute readers will note that the library is in the Advanced-Efficient zone. I included it in the general discussion of the three non-materials storage categories to keep continuity with that thought. Remember, this is all a connected offshoot of the Optimized Physical Layout element. At the beginning of your journey with the Storeroom Roadmap, there was a good chance that you didn't have a formal, well-stocked and well set-up technical library in your storeroom. During the organizational element of optimizing your layout, give this idea careful consideration and ask for and justify the resources needed for a great technical library.

BENCH STOCK

In addition to the three elements that make up the category of non-materials storage, there are other elements for consideration in optimizing the physical layout of a storeroom. One of those is Bench Stock.

You might find it interesting that Bench Stock is represented on the map as a strategic practice. The decision to classify, store, and issue an item as a bench stock item should be a conversation that is had by the representatives of the SSC. Further, what is stocked and maintained as bench stock should be evaluated a few times a year to make sure that items are moving, and the stock is turning. As bench stock is typically a vendor-managed stock and is often out of sight, out of mind, it would not be unheard of to have some unscrupulous vendors recommending that you stock all sorts of unneeded items. This should be audited occasionally. I would recommend it at least once or twice per year.

The term "bench stock" might be a local term, as I've heard it referred to by many names, "open stock" and "free issue" to name but two. I had the opportunity to learn the craft of maintenance consulting from some of the true legends in the field. One old, crusty consultant whom I admired greatly would get so mad if someone referred to these small, consumable type items as "free issue." In keeping with his southern roots, he'd turn beet red and yell, "Ain't nothin' free!" He must have had a bad experience with this type of inventory.

Items that are classified in this category are always low-value, high-quantity, and usually high-turning items. You know the parts: nuts, bolts, pipe fittings, etc. These are the items that are almost always out in front of the storeroom service window, and they are almost always exclusively serviced by a vendor-managed inventory (VMI) agreement (to be discussed later). You have this stuff. Look back again at the storeroom layout in Figure 2-2. The bench stock area is the space to the left of the roll-top door and directly on the other side of the service window. It is accessible through the personnel door, and the cabinets in the drawing are represented by black rectangles. Nothing fancy. It's just the nut and bolt bin.

There are some key concepts that help to make the bench stock area and service more efficient:

- Bench stock items are readily available in convenient locations throughout the facility; for example, bench stock may be in the maintenance shop, inside outbuildings around the facility, or on upper levels throughout the site (point of use).
- Bench stock is often located outside the physical confines of the areas under the jurisdiction of the storeroom; for example, nut and bolt bins in the maintenance shop occupy space in the *maintenance shop, not the storeroom*. But the bins and the inventory *are* part of the jurisdiction of the storeroom.
- Bench stock can be acquired without going through the formal Issuing process.

- Bench stock items are not usually kitted for planned jobs unless specifically requested.
- Where practical, bench stock will be part of a VMI agreement.
- VMI items usually have no part number, are not tracked by the storeroom, and do not factor into the Cycle Counting process.
- Bench stock items are usually charged against the GL (general ledger) account of the primary user. For example, nuts and bolts are charged against the maintenance department.

I recommend that the makeup of bench stock items be discussed and determined by the Stores Stock Committee. Remember that the production and safety departments are represented on the SSC. Some items available for bench stock can be production and safety supplies. Also, purchasing or procurement is also represented on the SSC. This department helps to coordinate the service contracts that will be necessary for bench stock operation.

The lighting is the last of these trailing elements to optimize the storeroom space.

LIGHTING SURVEY

I have been in more than one storeroom where the maintenance technicians had to activate their headlamps to read the fine print on boxes. Does this sound like the best that we can do?

A well-lit storeroom goes a long way toward having a safe storeroom. Associates are likely to see items blocking their path and are then able to work around obstacles. A well-lit storeroom is also a storeroom where people are likely to find the right part and record the part right. Lighting in the storeroom might not seem that important at 1:00 in the afternoon, but I believe it is at 1:00 in the morning.

The Office of Safety and Health Administration (OSHA) does not specifically have a classification for minimal lighting in a maintenance storeroom, but for warehouses and warehousing options, it specifies 10 foot-candles.

A foot-candle is a hard concept to get my head around. It might be easy for you, but I have spent a lifetime trying to figure it out. This is as clear a definition as I've ever read: "A foot-candle is the amount of light generated by a single candle that falls on a single square foot of surface no more than a foot away from the candle" (Schillaci). Couldn't be clearer.

This element is a bigger issue than initially thought. Lighting surveys are required once per year. The goal is to light every corner of the storeroom where someone might work or might be traveling. There isn't a mandate that the lights be on all the time, but the area the people are in at the time needs to meet the minimum requirements.

Organizations spent a lot of time and money in the early and mid-2000s relamping to LED lighting and other low-wattage options in an effort to save money. These projects had the added benefit of reducing the amount of heat being generated in the lighted areas. As noted earlier in the chapter, the last job I had as an employee was in the meat industry. I thought it was ironic that we spent so much money cooling the food areas at the same time we were heating them up with the old halogen lighting systems. Relamping was not a company priority at the time.

Lighting can also do a lot to lift the mood of the associates. No one likes to work in a dark and dingy workspace. I would encourage storerooms to consider painting walls and ceilings a light color to reflect the natural and the artificial light. Anything to help to see and read the small numbers that seem to be required reading in storerooms worldwide.

In my practice and in my classes, I use a free lux meter available on all smartphones. It doesn't matter what light meter you download and use; consider doing so in order to test your light level at any given time.

I'd suggest that the annual lighting survey be conducted by an electrical service. I'm certain their light meter devices are calibrated and traceable to a national standard. The free smartphone application is just a quick reference and should not be considered something you could or should take to court.

There is one final section to this long and detailed zone. The category is managing value, with a lead-in from the Last Used Report.

LAST USED REPORT

The conversion of the Last Used Report from a KPI to a tactical process is the most radical change to the Storeroom Roadmap I've made over the years. I kept wrestling with the idea that more aggressive attention needed to be given to the movement of goods through the storeroom. For all the reasons of controlling, minimizing, evaluation stratification, and care provided for spare parts, it just sat on my brain that storerooms should constantly be monitoring the issue history of the items in their charge. As a result of exercising this element, it would be a good idea for an efficient storeroom to publish (through the SSC) a list of items that might be nearing their consideration as an obsolescence candidate.

You may note the dashed line connecting the Last Used Report to the Optimized Physical Layout element. This is an *intended consequence* (not unintended but *intended*) and one that might come from a maintenance organization that is also on a continuous improvement continuum. If the maintenance and reliability effort is mature enough to start seeing an increase in component life on production assets, then there should be a corresponding decrease in parts usage. So much so that a part that used to be issued two or three times a year might not be issued again for two years. This could positively affect the size of a storeroom.

The purpose of the Last Used Report is to aid in the management of the inventory value and to enhance the inventory turn ratio (a KPI to be discussed later). The Last Used Report is a tool to identify potential candidates for obsolescence based on their last use date. In this instance, the last date of use is synonymous with the last issue date. This report is meant to increase the likelihood that no items will remain in stock indefinitely, having little or no issue activity, without some level of scrutiny.

Only items that are physically stocked in the storeroom are evaluated under the Last Used Report. This review does not apply to items that are stocked as "Order on Request" or otherwise "stocked" with no actual on-hand inventory. The intent is to manage the inventory value by evaluating the need for the on-hand inventory of obviously slow-moving items.

Likewise, this report is not applicable to those items officially identified as critical spare parts, as there is little desire to turn or use critical spare parts based on the implication of their use (e.g., failure in service is disruptive to production, and replacements have typically long lead times and are usually expensive).

Examples of items that would potentially be evaluated using the Last Used Report include:

- Routine spare parts
- Production supplies held in inventory in the storeroom
- HSSE supplies held in inventory in the storeroom

The Last Used Report will screen item activity and specifically target for reporting those items that have had no issue activity from stores for

400 days or more (as an example). Again, this is only a report of slow-moving items that are possible candidates for consideration for obsolescence, as previously mentioned.

The Last Used Report will report the following information on all on-hand stocked items (excluding critical spares):

- Date of last issue
- Days since last issue
- Part number
- Part description
- Where used
- Quantity on hand
- Cost per unit
- Total inventory value (of that item)
- ABC Classification of that item

The Last Used Report is run monthly (making it a tactical process) by the storeroom supervisor, and the outcome of the monthly report is forwarded to the primary users (departments) of those items for their review. If an item cannot be made obsolete entirely, the primary using department will be asked to consider if the on-hand inventory number can be reduced (optimized).

For example, if 20 bearings of certain specifications are on hand, and none have been issued for 400 days or more, could the on-hand count be reduced to 15 or 10? Can the item be eliminated altogether? Where the item is used and the number of items to be used at a single time (e.g., two belts are always used for a two-belt sheave pack) are key questions to consider when evaluating for reduced on-hand quantities.

In Chapter 2, we discussed roles and responsibilities. In fact, Table 2-3 covered this topic. You are now at a point on the Storeroom Roadmap that is a more mature location than you were in Chapter 2. Make sure that an associate is provided with the training and the motivation to keep this report current. This single report will be the pivot point in keeping obsolete materials to a minimum and will aid greatly in keeping the inventory count down per individual line item. You and your team should work to stay on top of this. The Last Used Report is often communicated as a standing agenda item at the monthly Stores Stock Committee meetings.

Table 2-2 mentions that the SMRP recommended best practice for inactive stock is to maintain a level at less than 1%.

This is a great lead into the managing value category that completes the Intermediate-Efficient zone. This category has six elements:

1. Obsolescence Criteria
2. Obsolescence
3. Item Substitution
4. Repair or Replace
5. Salvage Value
6. Disposal of Scrap

OBSOLESCENCE CRITERIA

Obsolescence Criteria is the companion element to Obsolescence. In this process, the SSC determines what makes a part or material obsolete. For the storeroom's purposes, an item is to be considered obsolete if it is no longer needed in support of an active asset or if the item is stocked at such a great level that there is no reason to believe the inventory will possibly be fully used. Simply stated, that item is no longer needed or is stocked at a level in excess of what will ever be used.

Items become obsolete for various reasons and can do so at any given time. As a result, there must be clear instructions on how to identify potential candidates for consideration.

Items become candidates for consideration for obsolescence when they:

- Are on-hand stocked items that have no recorded issues from stores for great than 400 days (for example)
- Are verified to be superseded by the end user (should be accompanied by a Management of Change notification)
- Are verified to be superseded by the storeroom (through contact with vendors and others)
- Are verified to be superseded by engineering and maintenance (through contact with OEMs, vendors, and others)
- Are only listed in support of assets that no longer exist

Four hundred days is a significant number; longer than a year, but short enough to avoid items getting lost in the system and not moving for years at a time. Identifying and reducing slow-moving or nonmoving inventory is a critical contributor to controlling stock inventory value.

If the storeroom and its partners in maintenance don't control the stock inventory value, then an agency that might not have the storeroom's best interest at heart will be forced to do it. Some of you have been there, and I'm guessing it wasn't enjoyable.

Another reason to control inventory scale and cost is the penalty of paying for carrying costs. Carrying costs for inventory average 18–35%, and the cost of carrying slow or dead inventory can be considerable.

An item identified as a candidate for obsolescence is not automatically removed from inventory. *Candidacy* only indicates that the item needs to be discussed regarding its continued value and current inventory level.

A scenario might indicate that a particular item (on-hand quantity of 20) has not had an issue for 405 days. This item is a candidate for consideration, and discussion will commence on its value and the value of the 20 units on hand.

Another scenario might indicate that procurement reports show that a vendor has recently discovered a problem with an item it provides and recommends that any inventory on hand be removed and discarded, and another product be substituted. This item is also a candidate for consideration.

The Stores Stock Committee should be consulted to determine the criteria for considering an item to be obsolete. I've used 400 days or more of non-use in my examples. That's just an example. The only way someone in the storeroom might know if a part should be considered for candidacy is to be alerted that it hasn't *moved* in a certain number of days. I guess the other method would be to monitor the thickness of dust on the items, but I would have a difficult time recommending that approach.

Obsolescence consideration will be given to any item received, tracked, and controlled by the storeroom including parts, materials, and consumables related to plant maintenance and operations, which merit the status. This applies to on-hand stocked items. This process is not intended for items that have no part number.

Without a proper means to identify components and materials that don't support an asset in the facility anymore, a storeroom could inadvertently end up stocking items forever. Some estimates place obsolete stock levels at anywhere from 58% to 60%. Imagine having an inventory valued at $2.7 million and finding out that 50% doesn't go to anything in the factory. That would be very poor management. Instead of keeping hold of parts that could be made to fit some application, why don't we take those out and buy the parts that truly do fit the application?

The Stores Stock Committee is the group that determines what it means for an item to be obsolete. Typical definitions are those that I've already written. One of the problems we have, and it's a bias, is coming to grips with the term "obsolete."

If you reported to the plant manager that your team had identified thousands of dollars of obsolete stocked items, the boss will likely think you did something wrong. "Why did you order thousands of dollars of obsolete inventory?" Items become obsolete for many reasons. Several of those have already been discussed. You want to have a process for identifying a part that might be a candidate for consideration.

The number "400 days" has been mentioned many times in this section. That is an unusual length of time and, quite honestly, was just made up.

Years ago, while I was working with a Stores Stock Committee on the Obsolescence process, I mentioned that I had gotten down to considering parts as candidates when they hadn't moved in a year (365 days). My client thought that one year couldn't possibly be long enough. I said, "How about 400 days?" He liked that better.

Some colleagues teach and suggest considering parts as candidates for obsolescence only when they haven't moved for three or five years.

A few paragraphs ago, I commented about an inventory value of $2.7 million and what if 50% was obsolete. Combine that thought with the earlier point that carrying cost could be 18–35% and you could be looking at some real business expense.

If your inventory value was $2.7 million and half of that didn't move in a long time and could be a candidate for obsolescence, then that would be $1.35 million in suspect parts. If your carrying costs were on the low end, say 18%, then you would realize a loss of $243,000 a year just for the privilege of stocking obsolete parts ($1.35 million x 0.18). At what point is it ridiculous?

One of the items your Stores Stock Committee will determine, almost as a first exercise, is to quantify what a critical spare part is and what an obsolete spare part is. Every other type of part is somewhere in the middle.

OBSOLESCENCE

The Obsolescence process naturally follows the previous element where the Stores Stock Committee established the criteria for obsolescence. The

Obsolescence process is used to evaluate all items stocked by the storeroom, including parts, materials, and consumables related to asset maintenance. This process is to be used for the continuous upkeep of inventory to ensure that only what is needed (in the quantity needed) in support of on-site equipment is maintained in the active inventory. This process will reduce storeroom inventory and lead to greater service. In other words, this process will make the storeroom more efficient.

Every storeroom should have as its primary goal, to have the right part at the right time and in the right quantity. Stocking items that don't support equipment flies in the face of the storeroom's prime directive.

Figure 5-23 is a good example of how the Obsolescence process would be initiated. One more time, notice the engagement of the SSC. This group is a hands-on committee that initiates and is therefore aware of most major efforts taking place in the storeroom.

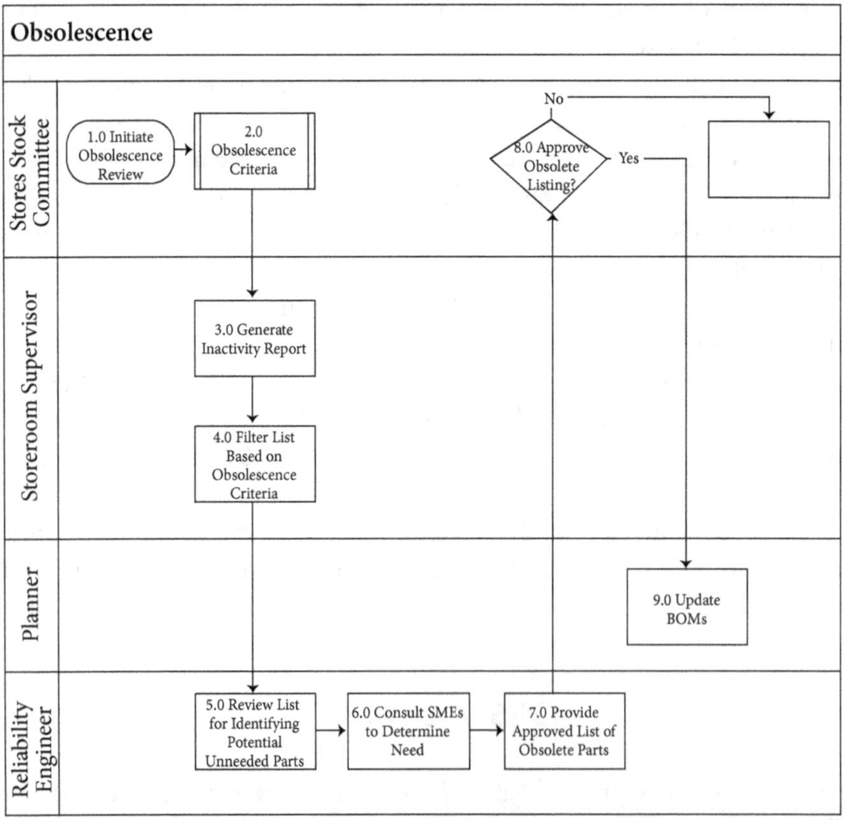

FIGURE 5-23 Obsolescence (partial).

I have been in the maintenance and reliability profession for almost 40 years. One of the most common grievances I hear in my travels and work within all industries is maintenance technicians complaining that "the storeroom just throws everything away." I think it is more likely that the storeroom has requested that the maintenance department send someone over to evaluate the continued need for some items that don't seem to have any movement. The maintenance department fails to respond, even though there may be many additional requests. Does the storeroom throw parts away without first trying to work with the technicians? It is doubtful. I've never seen it in almost four decades.

There is a counterargument against getting rid of obsolete items. It's important as a leader that you understand both sides of the argument. When items are identified as being in excess of what will ever be used in your facility, or the items are verified to no longer be needed in support of the assets on-site, is there any real harm to keeping them around?

"After all," a technician might remark, "we've already paid for the item. What is the harm in just leaving it here in case we could use it?" That is a legitimate question and one that we'd be fools not to address.

The fact is, whatever parts were purchased in an effort to cover all risk possibilities have already been paid for. Another hard truth is that it is unlikely that you will get your money back on the part. Vendors typically charge a restocking fee. Unless your procurement department is able to make a favorable deal with your vendor, you most likely will not get all your money refunded. And to be honest, some parts can be modified to work in a crunch . . . just in case. What's the harm in keeping the parts? If you can't answer this question, don't expect your technicians to jump at the chance to toss stuff from the storeroom.

The argument for weeding out unnecessary stock is primarily fiscal, but there should be an element that appeases the WIIFM (what's in it for me) mindset of those who use the storeroom's inventory; you know, the customer (maintenance).

We have already made a fiscal case against stocking unneeded items. We made some calculations and discovered that the hidden price of carrying cost would be too great. What harm is there in having a high carrying cost? It's not like the company is having to write a check for that amount each year. Is it even *real* money?

Carrying costs are an amalgamation of many costs:

- The cost of warehouse space
- The cost of labor to manage the inventory
- The cost of electricity and/or natural gas for the utilities consumed in stores
- The cost of insuring for the loss of the inventory
- The tax liability
- The loss of the value and use of money

That last bullet may be the real fiscal loss. Storeroom inventory is company capital. The equivalent dollar value of your inventory represents your company's investment. The company feels that it can get more value out of that sum of cash by stocking the right part and in the right quantity rather than purchasing another asset that would manufacture more products. It's a gamble the company is making. Let that sink in for a moment. Are you and the maintenance department leadership respecting the fact that the company chose to invest in you rather than another asset? The company is betting on you to provide a greater fiscal return than another machine. That's giving you and your colleagues some serious credit.

Your company might be better ahead financially if it would eliminate the storeroom inventory altogether and take that money and buy more production equipment. Who knows? This might already have been a conversation around your corporate boardroom. If it hasn't been considered, your dedication to running an effective and efficient storeroom will keep that thought from creeping into the boardroom to begin with.

In his best-selling book *Rich Dad Poor Dad*, Robert Kiyosaki starts a series of illustrations outlining the relationship between assets and liabilities (balance sheet) and income and expenses (profit and loss) (p. 62). Figure 5-24 is a re-creation of one of those illustrations in which I've superimposed three degrees of inventory management.

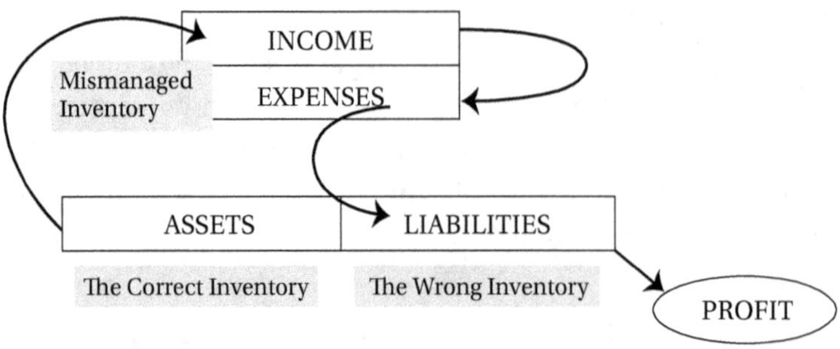

FIGURE 5-24 Degrees of inventory.

From Figure 5-24, we can see that Kiyosaki is telling us that *assets* make *income*. From that income, companies pay their *expenses* and service their *liabilities*, and the remainder is *profit*. I've superimposed three degrees of inventory:

1. The correct inventory is an asset.
2. Mismanaged inventory is an expense.
3. The wrong inventory is a liability.

Is an abundance of obsolete parts in the storeroom an asset to your company? No it is not. At best, it's an expense; at worse, it's a liability.

I was the maintenance manager for a steel mill for many years. My boss, Gilbert, was the superintendent of the melt shop. Our department was responsible for melting scrap steel and casting it into billets that were then sent to our rolling mill. We melted scrap steel with two 300-ton electric arc furnaces (EAFs).

Gilbert called me into his office one day to talk about something, but the conversation turned to production. "Did you know," Gilbert asked, "that we make 3,000 grades of steel here? Get your head around that for a minute. We make 3,000 different chemical compositions of steel at this plant!"

"I didn't know that," I said.

"Yes," Gilbert continued, "and we make some of those on purpose."

As it turns out, our little steel mill only sold 10 grades of steel. Over time, we made 2,990 heats of steel that were off-spec in regard to what we could sell. Gilbert went on to tell me that he had to contract with a service company for a "guy with a cutting torch" to come out and spend a year cutting up the billets so we could spend the electricity to melt them down again.

Did those 2,990 batches of steel billets represent mismanaged inventory or the wrong inventory?

What possible good can come to your company from having the wrong inventory? At some point, in some overt or covert manner, your company loses by having the wrong inventory at the right time.

The real struggle that most have with arguing against keeping obsolete material around just in case it's needed is that much of our inventory can be modified to be used. Some items have utilitarian benefits; after all, the item is already paid for. Who hasn't used a paper towel in lieu of a napkin, or a tissue in lieu of a bandage?

This is an absolutely true exchange. During the Middle Eastern conflict known as the Gulf War, I was sitting in my tent on our base in Taif, Saudi Arabia. My tent-mate had cut his finger on something and asked if I had a Band-Aid. "No," I said, "but I have a sandwich." Whatever look you have on your face right now is the look my tent-mate gave me. The wrong item is the wrong item.

Still, our technicians are quite the craftspeople, and if anybody can make a miracle happen, they can. In the Air Force, I had an old crew chief who would say, "Bad hydraulic fluid is better than no hydraulic fluid." The rest of that quote goes, "at 35,000 feet." On the ground, you don't want bad hydraulic fluid. In aircraft maintenance, you don't want any part or material that is not the correct item. Maintenance is entrusted with guarding the inherent reliability of the assets. How does MacGyvering a part to *make* it work contribute to the inherent reliability? It doesn't. In the airplane maintenance business, you don't want creative maintenance. For those of you in the pharmaceutical industry, you know what I mean.

"But John, what if it's the middle of the night, and we don't have the right part, but we have an old part lying around that we could make work . . . in a pinch?"

I've got a better idea. Why don't we start identifying parts that might be in excess of what we will ever use, and parts that don't go to anything anymore and get rid of those? Hopefully, we can get some money for them (see "Salvage Value" later). We will also free up some shelf space. With the money, or at least the reduction in inventory value, and the space we have freed up, we can buy the *actual* parts we need.

Technicians incorrectly assume that when we get rid of parts from the storeroom, we aren't allowed to buy the real parts we need.

The WIIFM question that maintenance technicians rightfully have is a simple one to respond to: "Do you believe that a world-class storeroom has the right inventory to maintain the inherent reliability of company assets, or do you believe that storerooms should stock stuff that 'might' work?" If that doesn't work, ask this: "Does a world-class technician deserve to be supported by a world-class storeroom?" Also, don't make the classic mistake of improving the storeroom performance without an equal improvement happening in the maintenance department. The truth is, you should never need a part in the middle of the night.

Years ago, I was conducting a 5S storeroom workshop at one of the customer fulfillment centers of one of those big parts suppliers (think

Grainger, only it wasn't Grainger). The facility was 2 million square feet. The 14-person maintenance team was responsible for the care and maintenance of the facility as well as the incredible amount of conveyance systems on-site (much of it was automated, high-tech equipment). This facility had a 300-square-foot storeroom, but no storeroom attendant. The maintenance folks stocked whatever they wanted, and it was as much of a get-what-you-want place as you can imagine. An interesting aspect of this site was that practically any item a maintenance person might want was a line item in the fulfillment center. Do you want a 12 × 24 × 1 pleated air filter? Aisle 19. Want a10-mm socket? Aisle 27.

The first S in 5S is "sort." For storerooms, *sorting* means going through the items in the storeroom space and eliminating what is not needed for the activities required. Obsolete and unusable items are not needed and should, therefore, be *sorted* out.

Since there were no storeroom attendants at this particular site, my 5S team was made up of the maintenance technicians. I had all 14 on my team. If anyone was qualified to go through the storeroom inventory and determine what was no longer needed, it was these guys. In one day, my group of SMEs (subject-matter experts) identified 2,000 pounds (yes, 1 ton) of stocked items verified to be of no use to the care and upkeep of the facility or the equipment. One ton! During the rest of the workshop, this team identified another 1,000 pounds of items that were more than what would ever be used.

I tell my classes that the 300-square-foot storeroom rose 3 inches from the ground for all the weight we took off it. That was an extreme eye-opener.

Recently, a client told me that his organization had the practice of stripping down equipment as it was heading out the door for the junk pile. The employees would strip any component they felt would and could be of use in the future. I smiled and said that I had done the very same thing myself. I suppose many of you have done this as well. I used to do that until I realized that the life of those components had been compromised by their years of service and whatever abuse the parts took while out in production. I also wondered if our company was so strapped for cash and had such poor regard for reliability that we would reuse a limit switch from an old machine we were trashing. Where does this end?

Here is my final thought on obsolete items and this element that makes our storerooms more efficient. The storeroom will not stock parts that don't support active assets in the facility. The storeroom will not stock partial parts (where a part was taken off a part) or used parts unless they are formally and certifiably rebuilt. We just won't.

As noted before, the maintenance department is responsible to communicate to the storeroom:

- What to stock
- In what quantity to stock it
- What to get rid of

As this particular category in the Intermediate-Efficient zone covers processes and practices that are used to manage value, there are other elements to consider in this category.

ITEM SUBSTITUTION

In the previous section, I talked about the poor practice of keeping obsolete items around to use in a crunch. That is just a poor approach to inventory management and just a poor maintenance practice. It's OK and even necessary to make something work, *anything*, in a real crisis. But effective maintenance leadership and sound reliability practices should have you working to distance your organization from experiencing any crisis situation. One area that can be exploited to make formal and approved options available is Item Substitution.

The word "substitution" can have many interpretations.

A substitution could mean superseding one item by another. Substitutions can be complete substitutions, where one item is completely replaced by another, or can be partial, in the case where the new item is only used for certain assets, and the original item was used on some original assets.

A good example of this type of substitution is one that most are familiar with when cooking. Vegetarians prepare many dishes substituting plant-based protein for animal-based protein. This is a good example of a substitution where one item is superseded by another.

Another definition of substitution is the temporary (or permanent) use of an item for the one that is actually required. I've encountered this many times at an auto parts store. My local auto parts store might not carry my exact brand or part, but there is a Ford-approved substitution.

The part that is being substituted for the desired part is an approved and useful part. It is, itself, a stocked item. It is not a part that is stocked solely for the purpose of being a substitute part; it is a part that is also attached to an active asset in the facility.

I hope you can sense the subtle difference between keeping an obsolete part around just in case it can be modified to meet some need vs. having an actual, viable spare part (that goes to something else) that can be modified, or used, to support a part that is needed but not on hand. With one, the part has no other use than to be modified. For the other, it truly goes to something else.

There is a field in the Item Master Data that is used to indicate that a part can be used as a substitute part in place of another. I don't believe this exists in all data tables, but you might look to see if your inventory management system has this feature.

Consider this scenario.

A major production line at the Atlas Company suffers a component failure at 2:30 in the morning. After 10 minutes of troubleshooting, the maintenance technician notifies the production supervisor that a motor coupling has sheared. The production supervisor calls up the storeroom inventory listing on a nearby computer and finds that although the coupling is a stocked item, the inventory is zero. What can be done?

Thankfully, as a consequence of a good Data Scrubbing review, the item master data for the coupling is complete and up to date. There is another coupling that is shown to be a suitable substitute for the one that has failed. And that coupling shows an on-hand quantity.

I've combined a lot of best practices in this scenario. Until your processes in maintenance are more mature, your technician is likely to go to the storeroom right away and try and locate the coupling. Not finding any on the shelf, the technician might report to the young production supervisor that there is not a part available, and all sorts of crisis management takes over. The technician might be a very good technician and may have noticed that a few slots over the shelf are couplings that look like they might work, but it's not certain. But it might be worth a shot.

Which outcome for the scenario considers the MOC process at your location, if you happen to have one? Which outcome supports a reduction in crisis management and MacGyverism? Which outcome supports Tom Moriarty's definition of "proactive": "I determine the outcomes by influencing the future," mentioned earlier in this book?

At this element in the Storeroom Roadmap, you should be well beyond the less-than-ideal practices of a common maintenance organization. A mature and consistently well-run storeroom will positively affect the maintenance organization and "drag" it (if necessary) along a continuum of best practices. It just will.

There are some components in the storeroom that are repairable. Not all, but some. When these components are issued, the storeroom should receive the "old" item back in order to send it out for evaluation and repair.

REPAIR OR REPLACE

This process is necessary to handle items identified as repairable assets when they fail in service, and can be repaired. Some repairable items that fail in service are notably non-repairable due to extensive damage. This process describes the instances when that is true and who is responsible to gauge the practical sense in repairing an asset, and when not to.

Additionally, this process addresses who is responsible for establishing the list of repairable assets and vetting the repair facilities.

This element leans on the relationship between the storeroom and the maintenance planner. At some point, even the finance department is brought in for its expertise. The good news is that all these agencies are represented on the Stores Stock Committee.

Regardless of who has oversight of the storeroom, the maintenance department is responsible for identifying (you've seen this before):

- What to stock
- In what quantity to stock it
- What to get rid of
- What parts are repairable

If it wasn't made clear in earlier mentions of this list of responsibilities, the person, the actual person who is responsible for this communication, is the maintenance planner. Does it make sense that the maintenance planner is aware of everything that is in the storeroom? Because of that knowledge, the planner also knows what is not in the storeroom.

The planner is responsible for communicating to the storeroom what components are regarded as being *repairable*. By exclusion, a component is non-repairable if not identified as a repairable item.

A repairable item should always have a unique serial number for reference to identify that specific item. This is a very important detail, as the planner will want to track the performance of items that have been repaired to ensure that their service life is still within the range that is tolerable for your organization. If you are familiar with the reliability metric of MTBF (mean time between failures), the planner should track the life depletion (if any) resulting from multiple repairs.

Slater recommends using bar coding to "track individual items to check on service life and repair effectiveness" (p. 219).

Let Figure 5-25 be a quick reminder of the time span of MTBF.

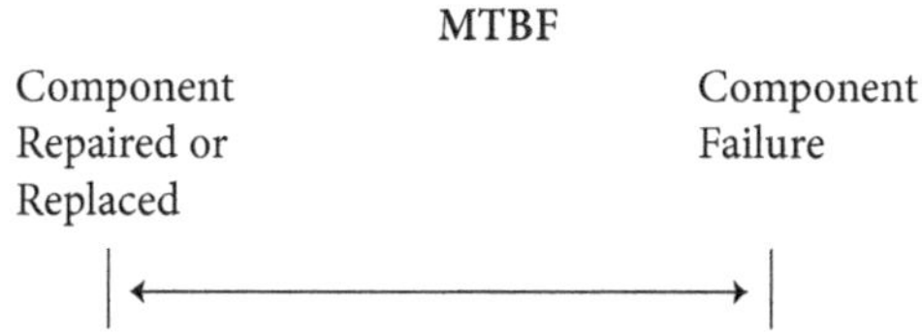

FIGURE 5-25 Mean time between failures.

It is common to see a decrease in life and usage for repaired items. A repaired car engine doesn't usually come with the same factory warranty as a new engine, nor should it have the same life expectancy. In Marshall Institute's eight-hour overview *Materials Management for Maintenance*, there is this notice: "Motors lose 3% efficiency with each rewind which translates to increased power consumption and shortened life" (p. 31).

I've read that the actual number is between 0.3% and 3%. Even at 0.3%, 10 rewinds on a motor will decrease its efficiency by 3%. If you are on the high end of that, you'll lose 30% efficiency. Why is this important? Motor efficiency goes into the calculation of torque. The function of an electric motor is to convert electric power into rotary torque. It doesn't take too many rewinds to affect the motor's ability to turn a load.

Figure 5-26 is a quick look at calculating torque for a motor.

$$\left[\frac{V \times A \times \%\ \text{E˜cien cy}}{746}\right] \times 5{,}252$$

FIGURE 5-26 Torque calculation.

This is not a book on reliability, but look at that formula one more time. It is unlikely that the power to the motor is exactly what you think it is.

Motors that are not aligned to 0.000 will have some adverse consequences, which usually appear as higher-amp draw and slower motor speed (RPM). There are enough uncertainties with electric motors in our facilities; losing efficiency by too many rewinds shouldn't be one of them.

For a general understanding of the workflow routing of a repair component, Figure 5-27 provides a good overview.

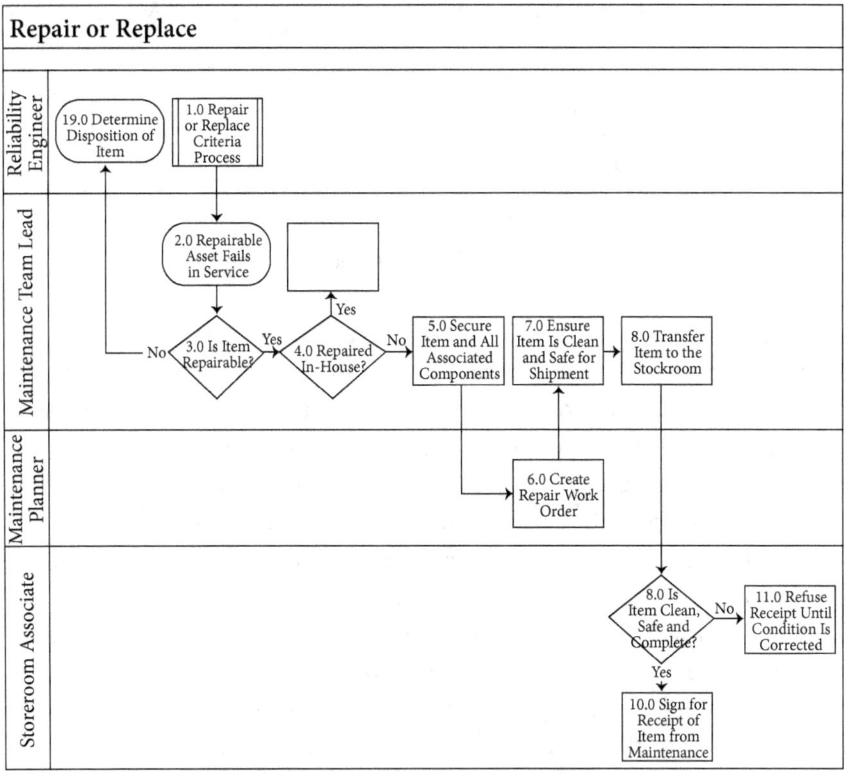

FIGURE 5-27 Repair or Replace (partial).

Although the planner is the person who communicates the status of a component as being repairable, it is usually the responsibility of the reliability engineer to make the determination. If you don't have a reliability engineer, then the maintenance leadership should designate roles and responsibilities for who should perform this function.

Once a component is identified as repairable, that component has an entry made in the Item Master Data signifying that it is repairable. By having this documentation in the single source of all item information, the storeroom associates, the planners, and the technicians all have the expectation that the process is going to be followed. Consistently.

Repairable components should also have a visual indicator, like a tag or some other "flagging" device to draw attention to the fact that they should be managed differently. Most notably, when the item is issued out, the technician tasked with the maintenance assignment is obligated to bring the failed unit back to the storeroom to be shipped out for repair or to be repaired in-house.

As a side note, I strongly encourage you to limit in-house repairs. Many of you are saddled with a requirement to extend in-house repairs as much as possible. I don't encourage this if at all possible because an in-house repair cannot provide the same money-back service guarantee that a contract service can. This might just be a personal preference.

When the technician brings the failed unit back to the storeroom, some courtesy rules apply. The failed unit must be:

- Clean
- Free of leaking fluids (drained)
- Safe (capacitors, sharp edges)
- Complete (no missing parts)
- And preferably strapped to a pallet if it's a large item

Items being sent out for repair should not simply be dropped at the doorstep of the storeroom.

Repairable components will be shipped with authorization to approved vendors. The maintenance leadership should receive an evaluation report and cost estimate. It is left to organizations to determine when a part is no longer fiscally feasible to repair. I've heard as much as 50% of a new item. If the cost of repair is 51% of a new item, buy a new item. I don't have a suggestion. Perhaps this is something for the SSC to vote on.

If the item is repaired and returned to the facility, the finance department or some division of your purchasing/procurement might be engaged in the discussion of what price to assign to the restocked item. Some companies stock the rebuilt item on the shelf at the price of the rebuild. Other inventory management systems assign the part an average price or average of the new and rebuilt motors. Again, this is a detail that must be determined. And it is a significant detail.

The rebuilt unit should be restocked with a trailing alphanumeric value at the end of the part number. I've seen an R used to indicate that that particular item, by serial number, is a repaired item.

All this is important to the planner when creating the pick list and generating the job plan for a repair assignment. The planner might indicate to the storeroom that a repaired component is not desired due to the sensitive nature of the particular repair. Also, for the cost estimate of the job, at what value is the repaired part sitting on the shelf? This detail was just discussed. A job could cost $1,500 or $10,000, depending on which component you want to use. That matters to a maintenance department trying to stick to a maintenance budget and trying to keep track of life-cycle costs.

By definition, and without exception, critical spare parts are repairable items. Unless the item is absolutely destroyed, I've never seen a one-and-done critical spare part. How could engineers expect to keep their job if they were to design equipment with highly production-threatening, costly, and hard-to-get parts that are disposable? That would be a bad combination.

All repairable components should be tested by your facility when they return to the plant. The repair shop undoubtedly provided a limited warranty. You don't want that warranty expiring while the untested part sits on your storeroom shelf for two years.

The first company I worked for after the military had an operation in the state of New York. I was given a tour of that plant in 1995. That maintenance organization had a dedicated PM/PdM team. One of the team's responsibilities was testing rebuilt components when they were returned from the repair shop. This PM/PdM team had a little shop and all sorts of jigs and fixtures to apply power to a device, run compressed air or hydraulic fluid through something, or even test circuit boards. I was very familiar with this practice, as similar test systems in the back shops supported aircraft maintenance.

If you don't have this type of testing and repair follow-through on the repairable parts, how would you know the part was ever repaired? How can you guarantee that it is fit and ready for service? Because the vendor told you so?

If your maintenance organization does not have this practice, when that repaired part is pulled out of stock after two years in your storeroom, you might be blamed for it not working at all, or for failing to last for very long.

There are two remaining practices in this roadmap zone. Both answer the same question: "Now what?"

SALVAGE VALUE

The Salvage Value strategic practice is probably the most utilized and least formalized practice in your organization and on this Storeroom Roadmap. This is the most proactive space on the map. This is the very point in a storeroom operation where you must know how to get rid of something. Items that you may need to dispose of might include but are not limited to:

- Obsolete items
- Slightly used items that you don't want to repurpose
- Work and project residue
- Items that are excess to you

This is literally the last stop where an item might sit while the organization determines how to get the most value from it.

This practice is part of the effort made by the storeroom to safely and properly dispose of materials and items that are confirmed to no longer have value for the operation and mission of your company. From time to time, and for various reasons, items and materials in stock or ordered for activities are deemed to have no value for the mission, yet those same materials and items may still have some monetary value to someone. This practice guides the decision-making for the proper disposition of those items.

Careful consideration should also be given to the question of liability to your company and what sales or donations could mean because of unintended consequences (e.g., you sell an item to a civilian and the person gets hurt by that item).

I mentioned that this is probably the most proactive element on the whole map. Look at Figure 5-28, the early process steps in the Salvage Value practice. I contend that an active and engaged Stores Stock Committee could easily map out the desired process for disposing of usable items that still have some value and for capitalizing on a financial return.

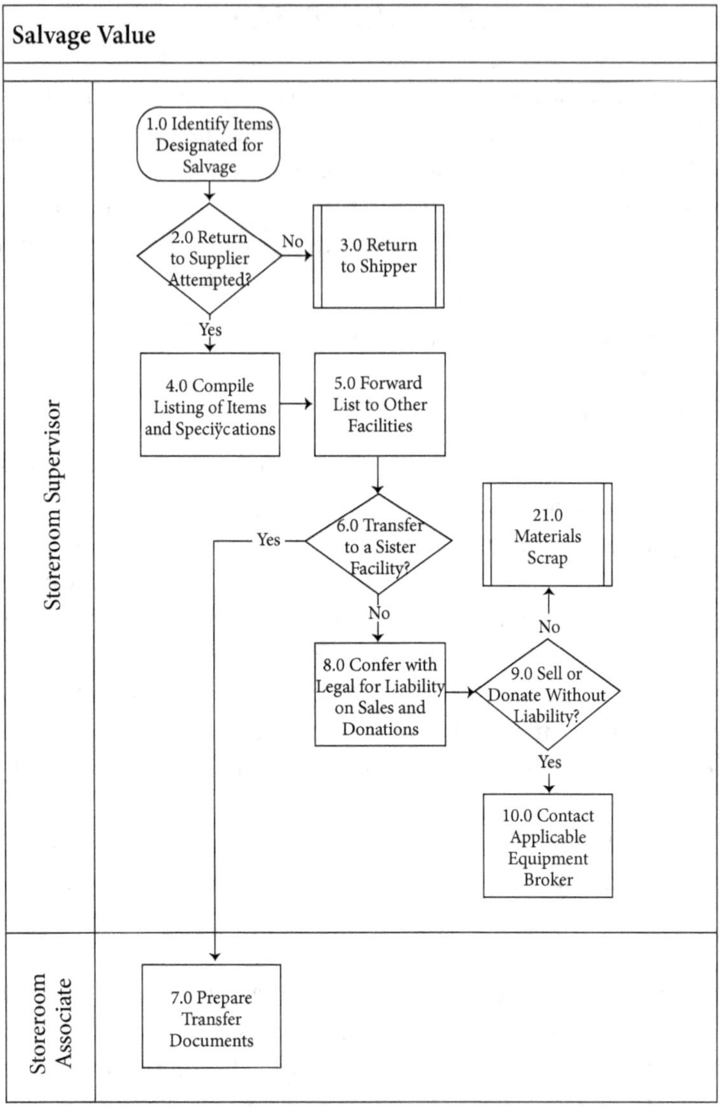

FIGURE 5-28 Salvage Value (partial).

If you stay with the theme of being proactive, can you see how parts standardization across your organization could help dispose of components that you do not need at your location? Other locations would have the same part and could make use of your surplus.

Also, an organized and optimized storeroom would likely be a storeroom where the original containers for the items are intact. We will learn more about preserving the functions of the parts in the advanced zones.

Finally, good relationships with vendors and suppliers could lead to their agreement to buy back inventory that has not been used and has been well cared for in stores. This was a principle of the Return to Supplier element.

You should have a comprehensive strategy to deal with surplus storeroom items and those deemed to be obsolete for your purposes. Why? Because you will always have the need to do so. Usually, the manner in which companies are able to recognize some financial compensation for getting rid of spare parts and materials they no longer need follows this order of preference:

- Sell them back to the supplier.
- Sell them to a sister company or facility.
- Sell them to a broker.
- Sell them through online sales.
- Donate them for tax credit.
- Recycle them.
- Scrap them.

In the end, however, if you can't sell or give away a component that is surplus to your facility's needs, you might need to scrap it.

DISPOSAL OF SCRAP

This element is the end of the road for the Intermediate-Efficient zone. It has been a long trip and the longest section of the Storeroom Roadmap. At the beginning of this chapter, I mentioned that this zone was akin to the mountaintops of our journey. From here, you can see the final frontier, the advanced zones. But we aren't done traveling through this zone just yet.

This final practice is vital for the efficient, safe, and environmentally responsible collection and disposal of materials scrap. Materials scrap is generated almost on a daily basis. The direction in this guideline recognizes that these materials might be work residue (e.g., conduit, pipe segments), but might also be obsolete stocked inventory that cannot be disposed of to gain any salvage value and is therefore slated for scrapping. This is where items that can no longer be repaired end up. There should be a final disposition for them.

Regardless of how the material is identified as scrap, materials that can be recycled and have been established as requiring proper collection need to be handled in a safe manner with a focus on the smooth transition between your company and the approved scrap dealer. It is common to see the recycle bins used by the maintenance department located by the storeroom and possibly even managed by the storeroom. This is never a desirable duty, but a necessary one.

Of course, Disposal of Scrap goes beyond the recycled materials that are common in most maintenance organizations. Often, batteries, paint, and other hazardous chemicals might fall under this practice.

Similar to Salvage Value, this practice benefits from a lot of preplanning and solid vendor relationships.

What are your other choices if an item cannot be sold or returned to the supplier if it can't be placed in a recycle bin due to a high concentration of some desired metal or material? Some facilities just threw the item out in the back forty a long time ago. That wasn't good for the environment, and it still isn't.

The least effective and the least desirable means of scrap disposal is to throw the item or material in the trash. Obviously, don't do this for hazardous items. I hate that it comes down to this sometimes because we don't need more industrial waste in landfills.

If you get to this point, throwing items in the trash, there were a lot of missed opportunities you had to navigate through, and you should evaluate all the elements up to this point to determine how the hell you got here. An honest assessment will clean up your act and clean up your scrap.

SUMMARY

The intermediate zones take up a lot of space on the roadmap. This is part of a very long road trip that requires patience, good planning, a steady hand at the wheel, and an exceptional navigator.

We began the Intermediate-Efficient zone by detailing the companion elements to those in the Intermediate-Effective zone. Right away it was made clear that the Stores Stock Committee would not be the usual coffee-drinking, doughnut-eating committee. No sir. This is a committee of engaged decision-makers with some decisions to make.

After making effective processes and practices more efficient, our journey took us over to one of the most popular sites on the map. Sorry, I can't help myself in making these map references. Getting stocking levels correct is always a highly anticipated spot on the roadmap. But this is the wrong place to start a journey. You might make a little progress, but it won't be sustainable. Just look at the map again, and notice all the organizational elements, KPIs, and processes and practices you must have solidly in place before you should even consider such a move.

Evaluating your stocking levels and your stocking methodologies is a practice that requires a lot of work from a lot of serious folks. You will find out that you don't need to stock everything; in fact, you don't want to. But you should know where to get just about everything.

We also discussed how to requisition items that aren't even stocked. Regardless of how items are ordered when they arrive in the storeroom, they are meant for stock replenishment or a job kit. No exceptions.

Some very large initiatives were introduced that could benefit corporations if done correctly. There is little argument that parts standardization goes a long way to improve efficiencies, as does a comprehensive bill of materials. Both of these rely on engineers doing their jobs to get these elements correct, right out of the gate.

An extended discussion was centered on re-evaluating the physical layout of the storeroom. An organization that takes this roadmap approach seriously will find itself, a few years into its journey, realizing that it needs to move some stuff around and take another look at the storeroom setup for future service and convenience. A great analogy is one you see all the time when families move because their immediate and future needs evolve. The storeroom inventory should be leaner, cleaner, and more in line with exactly what is needed to keep your crazy operation working. That sounds like a storeroom that is ready for a new look and smoother, sleeker layout. There are many ancillary categories of stocked items that also need to be given consideration.

The zone ended with a discussion of what to do with inventory that is slow-moving or just no longer needed to support the facility's effort. The last thing anyone wants to do is to throw a good spare part in the trash can.

WORKBOOK

The workbook for Chapter 5 is going to be understandably longer than those in the other chapters. A lot of ground was covered in this Intermediate-Efficient zone. If you read the chapter carefully, I suspect you know where your issues are. I also believe you found some areas where you are exercising some very good practices. Keep that going. Let the workbook be a guide to help you establish some goals to continually improve.

Here is the assignment for Chapter 5. Fill in Table 5-5 with a Y for yes or an N for no.

TABLE 5-5 Answer Y or N

	Y or N	N/A
Do you have formal process guides created for the following?		
Adjust ABC Classification Model		
Cycle Counting Criteria		
Optimizing Stock Levels		
Requisition of Materials		
Receiving Purchased Items from Stores		
Parts Standardization		
Special Tools		
Safety Items		
Bench Stock		
Obsolescence		
Item Substitution		
Salvage Value		
Disposal of Scrap		
Does your maintenance organization identify items that are repairable?		
Does your organization work to achieve parts standardization?		
Do you perform an annual lighting survey?		
Do you use a mathematical formula to calculate min/max?		
Do you use an algorithm to determine the criticality of a part?		
Do you cycle count a little bit everyday?		
Do you have a technical library that is in good shape and useful?		

Many short videos, listed in Table 5-6, cover this zone's many elements. Work through this library. I believe you will find that my message is consistent and well within the general thinking of best practices. Best of all, it works!

TABLE 5-6 Topic Links

Title	**Maintenance Innovators, Inc., Website Link**
Critical Spare Parts and Algorithm Review	https://maintenanceinnovators.com/aiovg_videos/critical-spare-parts-algorithm-and-review/
Optimizing Stock Levels	https://maintenanceinnovators.com/aiovg_videos/optimizing-stock-levels/
Min/Max vs. Safety Stock	https://maintenanceinnovators.com/aiovg_videos/min-max-or-safety-stock/
Standardization	https://maintenanceinnovators.com/aiovg_videos/standardization/
Special Tools	https://maintenanceinnovators.com/aiovg_videos/special-tools/
Last Used Report	https://maintenanceinnovators.com/aiovg_videos/last-used-report/
Obsolescence and Its Criteria	https://maintenanceinnovators.com/aiovg_videos/obsolescence-and-obsolescence-criteria/
Item Substitution	https://maintenanceinnovators.com/aiovg_videos/item-substitution/
Repair or Replace	https://maintenanceinnovators.com/aiovg_videos/repair-or-replace/
Salvage Value	https://maintenanceinnovators.com/aiovg_videos/salvage-value/
Disposal of Scrap	https://maintenanceinnovators.com/aiovg_videos/disposal-of-scrap/
The Storeroom as an Asset	https://maintenanceinnovators.com/aiovg_videos/the-storeroom-as-an-asset/
Bench Stock	https://maintenanceinnovators.com/aiovg_videos/bench-stock/
	YouTube Dr. John's Maintenance Minute Link
Critical Spare Parts and Algorithm Review	https://www.youtube.com/watch?v=UeT9EsOIArI
Optimizing Stock Levels	https://www.youtube.com/watch?v=rrFjOYzj3tw&t=17s
Min/Max vs. Safety Stock	https://www.youtube.com/watch?v=zbp-k9tGdrE&t=8s
Standardization	https://www.youtube.com/watch?v=qPPAoBvQlJE
Special Tools	https://www.youtube.com/watch?v=NfIXm0-cP54
Last Used Report	https://www.youtube.com/watch?v=rYzXSc5AqpM
Obsolescence and Its Criteria	https://www.youtube.com/watch?v=cLP-SEU9CqY&t=7s

(continued on next page)

TABLE 5-6 Topic Links (*continued*)

Title	**Maintenance Innovators, Inc., Website Link**
Item Substitution	https://www.youtube.com/watch?v=hJoyk3DcOPQ
Repair or Replace	https://www.youtube.com/watch?v=RZ3J6Pn0SFU&t=21s
Salvage Value	https://www.youtube.com/watch?v=PI278l8Ffmw
Disposal of Scrap	https://www.youtube.com/watch?v=PU999P5oxUU&t=19s
The Storeroom as an Asset	https://www.youtube.com/watch?v=nK9Dqt--9HM
Bench Stock	https://www.youtube.com/watch?v=Epb80tE45A0&t=39s

CHAPTER 6

ADVANCED-EFFECTIVE

Good judgement is the result of experience
and experience the result of bad judgement.

—Mark Twain

The good news? You've reached the other side of the mountains on the Storeroom Roadmap, and the valley is right in front of you. Smooth sailing ahead. The better news? There are only two elements in this zone. After the Intermediate-Efficient zone, I thought you might need a rest.

Figure 6-1 lays out the two elements in this zone. One is a rarely used or often improperly interpreted KPI. The other is the behemoth process you will utilize to keep your spare parts in good order.

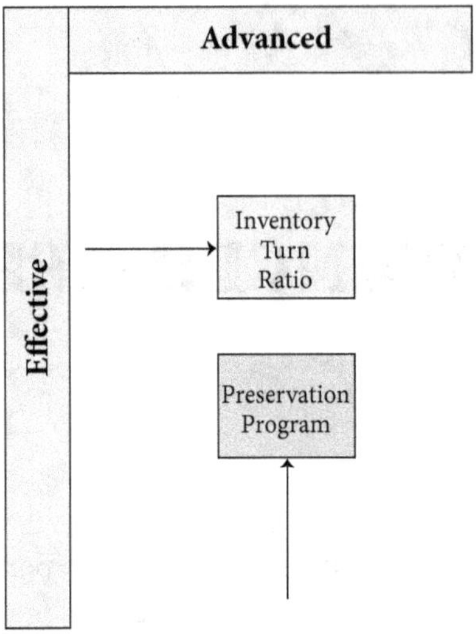

FIGURE 6-1 Advanced-Effective zone.

TABLE 6-1 Legend for the Advanced-Effective Zone

Element	Activity
Inventory Turn Ratio	KPI
Preservation Program	Process

Of the 32 processes and practices that are the central theme of this book, there is one that is titled "KPIs." The first element we encounter in this zone is the last metric we will discuss. The metrics discussed so far and this next one make up what I've come to name the "Big 3."

INVENTORY TURN RATIO

The Inventory Turn Ratio is the third of the Big 3. Sometimes "inventory turn ratio" is referred to as the "inventory turn rate"—they mean the same thing. The Society for Maintenance & Reliability Professionals calls this metric "stores inventory turns." SMRP states: "This metric is a measure of how quickly inventory is flowing through the storeroom inventory system. It can be applied to different categories of inventory, including spares and operating" (*SMRP Best Practices*, p. 255). Note that the definition from SMRP ends with the word "operating," but it seems to me that there should be another word after that. There is not.

SMRP further states: "This metric is used to measure the appropriateness of storeroom inventory levels" (*SMRP Best Practices*, p. 255). I dumbed it down to mean "a metric that calculates how effectively the working capital in stores is being used as opposed to being idle in bins or on shelves."

Although I've placed this metric in the Advanced-Effective zone, I should point out that other experts in storeroom management may consider this to be a metric meant for inventory management efficiency (Slater, p. 192).

There are some key concepts that are important to note:

- The turn ratio helps to analyze whether the correct items are being stocked, and stocked in the correct quantity.
- The turn ratio can be improved through the constant evaluation of the inventory, specifically looking for potentially obsolete or other slow-moving items.

In the introductory segment to this chapter, I mentioned that this is a metric that was rarely used or interpreted properly. I think it is rarely used *because* it is rarely interpreted properly.

The conflict and confusion come from the formula used and what a world-class or best-in-class value would be.

Figure 6-2 is the formula in SMRP's *Best Practices* (p. 255). I recommend that you use this formula if it makes sense and works for you. I personally have some questions, but SMRP's formula is internationally recognized as the recommended formula.

$$\frac{\text{Value of stock purchased over a set period of time}}{\text{Value of stock on-hand}}$$

FIGURE 6-2 SMRP's stores inventory turns formula.

In the formula from SMRP's *Best Practices*, the value of stock on-hand is the current value of the inventory you are holding in inventory. The value of stock purchased, and this is the first confusing point, is defined as "The value of the inventory items purchased in the period for which the metric is being calculated" (p. 255).

Figure 6-3 is the formula that I teach and use. There are some notable differences and points of conflict between my formula and SMRP's. I want to stress again, for the second time, if you find value and can use SMRP's formula, please use that one.

$$\frac{\text{Value of stocked items issued out of the storeroom in a set period}}{\text{Inventory value} - \textit{the value of critcal spare parts}}$$

FIGURE 6-3 Maintenance Innovators' turn ratio formula.

I want to address the differences in both the numerator and the denominator in both formulas. But first, a reminder. The turn ratio is going to essentially tell us if we are stocking *about* the right parts in *about* the right quantity. This metric alone can't be the only one you use to measure the storeroom's performance. If the turn ratio is about right, and you are keeping stockouts to a minimum and emergency purchases to a minimum, and if the service level is high, then you can certainly consider your operation to be effective and efficient.

The numerator. Table 6-2 is a summary of the two different numerators and their interpretations. Which one works better for you, and do you interpret the SMRP numerator the same way I do? It could be that you understand this value better than I do.

TABLE 6-2 Numerator Comparison

Numerator	Interpretation
Value of stock purchased over a set period of time	
	"Stock purchased" could mean stock items that were "purchased" from the storeroom, meaning only stocked items
	"Stock purchased" could mean all inventory purchased and received through the storeroom
	"Stock purchased" could also mean any item that wasn't even a stocked item, like a direct-ordered item
Value of stocked items issued out of the storeroom in a set period of time	
	The value of only the items that were stocked in inventory

The denominator. Table 6-3 is a summary of the two different denominators and their interpretations. Which one works better for you, and do you interpret the SMRP denominator the same way I do? It could be that you understand this value better than I do.

TABLE 6-3 Denominator Comparison

Denominator	Interpretation
Value of stock on hand	
	The value of the current inventory at the time of inquiry
Inventory value minus the value of critical spare parts	
	The value of the current inventory at the time of inquiry minus the current value of the critical spare parts

I can't speak for SMRP, but I can explain the meaning of the formula I have used for years.

As I understand it, and teach it, the turn ratio is a way of determining if what is being stocked is *turning* and moving out of the storeroom. "Turning," as in the inventory phrase meaning "turning over" and being replaced with fresh stock. In my class, I use the example of a store that sells only bread. Here is the scenario I share in class.

You own a grocery store that only sells loaves of bread. Each year, your store sells 120,000 loaves. If each loaf of bread costs you $1 and you sell 120,000 loaves of bread a year, you will pay $120,000 for bread in a year. To simplify the math, let's assume you sell 10,000 loaves of bread a month (12 × 10,000 = 120,000).

Scenario 1. You ordered 120,000 loaves of bread to be delivered on January 1. You put out some loaves and keep restocking the store shelves from your back warehouse. On January 1, you have $120,000 of your capital (money) tied up in bread. That's a lot of bread, in bread. That is a turn ratio for the year of $120,000 worth of bread sold over an inventory of $120,000, or $120,000/$120,000 = 1.

Scenario 2. You ordered 10,000 loaves of bread to be delivered on January 1. You put out some loaves and keep restocking the store shelves from your back warehouse until you get down to 2,000 loaves of bread. You order another 10,000 loaves of bread from your supplier at that time. At any given time, you will never have more than 12,000 loaves of bread, or $12,000 in capital, tied up in bread. That is a turn ratio for the year of $120,000 worth of bread sold over an inventory of $12,000 (roughly on hand at one time), or $120,000/$12,000 = 10.

Scenario 2 frees up $108,000 in capital. What could you do in your bread store with another $108,000? You could buy peanut butter and jelly to expand your store's offering.

That is what a favorable turn ratio can do for your bread store and your company's storeroom.

In my formula, for the numerator, I only consider the value of the items that are in the stocked inventory and issued out through the service window or in job kits as the value that goes on the top of this formula. Right away, that discounts the items that were ordered on request, as they were never physically stocked in the storeroom. It also discounts the value of the items not stocked or cataloged but ordered directly. Again, if the item is not physically stocked in the storeroom, the value of the item is not included in the numerator. The SMRP formula and definition didn't quite make their formula clear on this point.

My rationale is that I'm calculating the turn ratio of my *stocked* on-hand inventory. The *stuff* that is taking up space on my shelves

For the denominator, I think that SMRP forgot a detail. That detail centers on critical spare parts. If you recall from the previous discussions on critical spares, one shared characteristic of all critical spares is that you never want to use them. Replacements or repairs would be expensive and take a long time to get, and your organization would be at a higher risk level because you would be without a spare for a long time. For that reason, I elected to subtract the value of critical spare parts from the value of the storeroom's total inventory in the denominator.

Now, critics of my rationale might say, "What keeps us from making everything a critical spare part, thus driving the denominator down and the turn ratio up?" First, no one I know actually uses the word "thus." Second, the answer to that lies once again in the discussion of critical spare parts. I wrote, and you read, that you want to keep your critical spare parts down to a minimum, around 1–2% of your inventory SKU (stock keeping unit) count. And you are not going to label items as critical anymore; you are going to use a calculating methodology. Either use the one I gave you as an example, or create your own.

By not deducting the value of critical spare parts from the denominator, I feel that SMRP is almost forcing you to use your critical spares to have a reasonable turn ratio.

Next, let's explore the best-in-class value.

SMRP Best Practices has the best-in-class target value at >1 or >3 with critical spares (p. 257). I don't understand this value, as it is the first mention in this metric of special consideration for critical spares, and SMRP doesn't tell you if you should add or subtract the value of critical spares from the numerator or the denominator. I should have mentioned earlier, but I'll mention it here: This metric has no units because it is dollars divided by dollars, so the units cancel each other out.

The other questionable part of using this metric, and, by the way, you should definitely use this metric, is the target value. As I just reported, the SMRP values are >1 or >3 with critical spares. I've seen the world-class number reported to be 1.4 (which is admittedly greater than 1) for processing and discrete manufactures, with 3.0 being the value for chemical operations (MRM 1, p. 100). Other sources suggest that "the good performers will have a high level of stock turns" (Slater, p. 73). Doc Palmer writes, "Turns are the ratio of all inventory issued during a year to the cost of the inventory on hand. Between 1.5 and 2 turns is supposedly a good number" (p. 330). To top it off, Dr. Blache of the University of Tennessee states, "Best-practice annual inventory turns should be in the 2 to 4 range" (p. 6).

Everyone seems to be dancing around the same essential value, but is there a cap? Surely 100 turns a year is too much. Isn't it? Especially if the cost to issue a purchase order is $65.

Dr. Blache surveyed more than 200 facilities, and one of his questions centered on turn ratio. His study showed that many facilities operated at a turn ratio below 1: "The lowest value was 0.3, and the highest was 16, with an average of 2.4 MRP inventory turns. The data indicate that 46% of the facilities had MRO inventory turns less than 1, and 67% had MRO inventory turns less than 2. This further indicates that most facilities have too many spare parts" (p. 6).

Here is my advice and the counsel I give to my clients. I suggest they set a target range of 2.3–3.4 for their inventory turn ratio. Maintain this turn ratio until you have an exceptional preventive and predictive maintenance program and near best-in-class planning and scheduling. If you attempt to reach a turn ratio of 1.4 or 1 with ineffective PMs and planning and scheduling "in name only," you will suffer significantly in your storeroom's service and convenience. I predict you will need to fly everything in as an emergency purchase.

Why? Because a world-class storeroom has very little inventory. Its only stock is meant for true emergencies. Ideally, a world-class storeroom would have very little stocked inventory, with most needed items ordered directly to job kits for planned and scheduled work. With a turn ratio around 1 to 1.4, what stock world-class companies do have on hand would turn slowly.

In contrast, let me tell you about a location in Michigan I consulted at years ago. This storeroom stocked 1,600 different sizes of sprockets. You read that correctly. And it stocked 10 of each size. I was on a storeroom 5S assignment and was looking at a storeroom with 16,000 sprockets. If you are picturing shelves that are bowing under the weight, you have the image I had. Oh yes, one detail. The storeroom was on a mezzanine.

I asked the maintenance manager about the sprockets. "Surely," I said, "you don't have 1,600 different sprockets in this little plant?" He assured me that they did. I told him that even Spacely Sprockets didn't have 1,600 different sprockets. I'm not sure if anyone reading this got the reference, but Spacely Sprockets is the company George Jetson worked for in the 1960s cartoon series *The Jetsons*. Of course, my point was that not even a plant that makes sprockets has that many sprockets.

Looking back on what I've written for Inventory Turn Ratio, I realize I've been unnecessarily critical of SMRP. I didn't mean to be. I was only explaining why this metric is rarely used, and as I explained, I believe it's because it is misunderstood. There are many questions about the formula and the calculations.

One detail that I wholeheartedly agree with SMRP on is the need to break out MRO parts and materials from everything else. If you calculate an aggregate of storeroom items that turn, you are likely to sweep up operating supplies and safety items in that calculation. My intent here, and SMRP makes this point clear as well, is to measure the turn ratio of MRO parts and materials.

There should also be a further breakout of A items, B items, and so on.

I strongly suggest that you adopt this metric and the other two explained in this book—Inventory Accuracy and Inventory Service Level. These are the Big 3. The "Sergeant Majors," as I like to call them. Understand them, and determine the best way to measure and report them. Know what to do with the information. Make a plan, and get better. Slowly and methodically. Again, the three metrics detailed in this book are collectively referred to as "KPIs" on the list of the 32 storeroom processes and practices.

The second element, and the last one in the Advanced-Effective zone, is a big one. It is the Preservation Program.

PRESERVATION PROGRAM

I'm going to go out on a limb here and assume that if you are reading this, you have chosen to read the entire book first and are not waiting until this very moment to start a preventive maintenance program for your stored items. This element that rounds out the Advanced-Effective zone of the Storeroom Roadmap is where you take the time to *improve* the effectiveness of your stocked items' preventive scheme already in place.

As I'm often quoted as saying, assets come to your facility with what's called *inherent* reliability. If an asset were designed poorly, built cheaply, and installed hastily, you wouldn't have a world-class asset; you'd have a nightmare on your hands for the next 20 years. The role of the maintenance department, and indeed everyone, is to guard the inherent reliability of the asset. The asset's reliability is as good as it's going to get. You can't *maintain* more reliability into an asset than it has inherently.

Similarly, spare parts arrive at the storeroom with their inherent reliability. It is true that the very rare part might make it out of the manufacturer with some defect. But Six Sigma, quality circles, and total quality management (TQM) have pretty much reduced that possibility to near zero. It is more likely that the reliability of a spare part would have been compromised in shipment before it arrived at your receiving dock. Once it arrives at your location, whatever reliability a part has is the level of reliability you need to guard while the part is in the storeroom's care. These parts might not be pressed into service for months or years. Guarding the inherent reliability of the spare parts becomes the responsibility of the storeroom personnel in the process element of the Preservation Program.

Dr. Nathan Wright makes unambiguous links among an effective maintenance organization, asset reliability, and a properly and well-run storeroom in his book, *The Death of Reliability*. Dr. Wright notes:

> The third and final group is improper installation [Dr. Wright is discussing *contamination*]. This is made up of your maintenance, repairs, and operations (MRO), engineering, and your training. As I have repeated about the first two groups, this is a cost-effective way to be reliable. Inventory assurance is key to your MRO and the hub of your reliability. You are already spending money on the inventory, all you need to do is store it correctly. (p. 15)

There are two interesting takeaways from Dr. Wright's input. For one, his book credits the storeroom for being the hub of your reliability effort. I believe that is a point I've made in all three of my books. Well stated by Dr. Wright. Second, I thought it was interesting that he takes the position that mishandling and poor storage practices are a type of *contamination*. I'm not sure any other author has made this comparison, but I like it. I can certainly see where eliminating poor storage practices would be a type of *defect* elimination, but I never thought of it as a source of *contamination*. You and I both learned something here.

I will go into some detail about how to preserve parts in stock in this segment, but the purpose of this element is to improve the process procedures (the process guide) for how the storeroom is going to develop and execute the policies and practices that the company will elect to perform. There are clearly best practices that will be discussed a little later. In this zone, you want to make your current preservation program more effective and disciplined.

I'm taking the very rare step of publishing the entire workflow for the Preservation Program as I would recommend. Other workflows shown in this book are partial flows to show you some of the ideal steps your workflows might cover. This one is very important, and I want you to have what I consider a very good approach. Figure 6-4 is the Preservation Program workflow.

The Preservation Program executed by the storeroom is part of a larger preservation effort. The Preservation Program is meant to establish and deliver the necessary steps, tasks, and detailed procedures to protect and preserve items in storage. The environment in which the plant operates and in which the materials are stored dictates that a comprehensive program be developed and executed. The way an item is stocked and preserved in the storeroom will directly correspond to the expected life of that part in service. This life, of course, can also be compromised by maintenance installation deficiencies and abuse from operators.

The overall program will have activities to be conducted by maintenance personnel and those meant to be completed by storeroom personnel.

For example, highly technical tasks on variable-speed drives might specify that a particular skilled trade is needed to complete the PM, whereas tasks on rotating belts and bearings in routine stock would fall on a qualified storeroom associate.

This process lays out the steps, from a storeroom perspective, that allow for new items to be identified as needing a preservation process, contacting the necessary group to put those processes together, executing the PMs, and collecting data for reporting and problem identification.

Placing this process in a zone on the map that comes so late in your journey is a practical and viable location. Consider that the journey to this point began in the Foundation-Effective zone during the New Item Setup process. You can trace a path on the map connecting these two very critical elements.

I counsel companies worldwide on how to create an effective PM program for their spare parts. At first, this seems like a daunting task, but once you start developing the action steps, it will become easier.

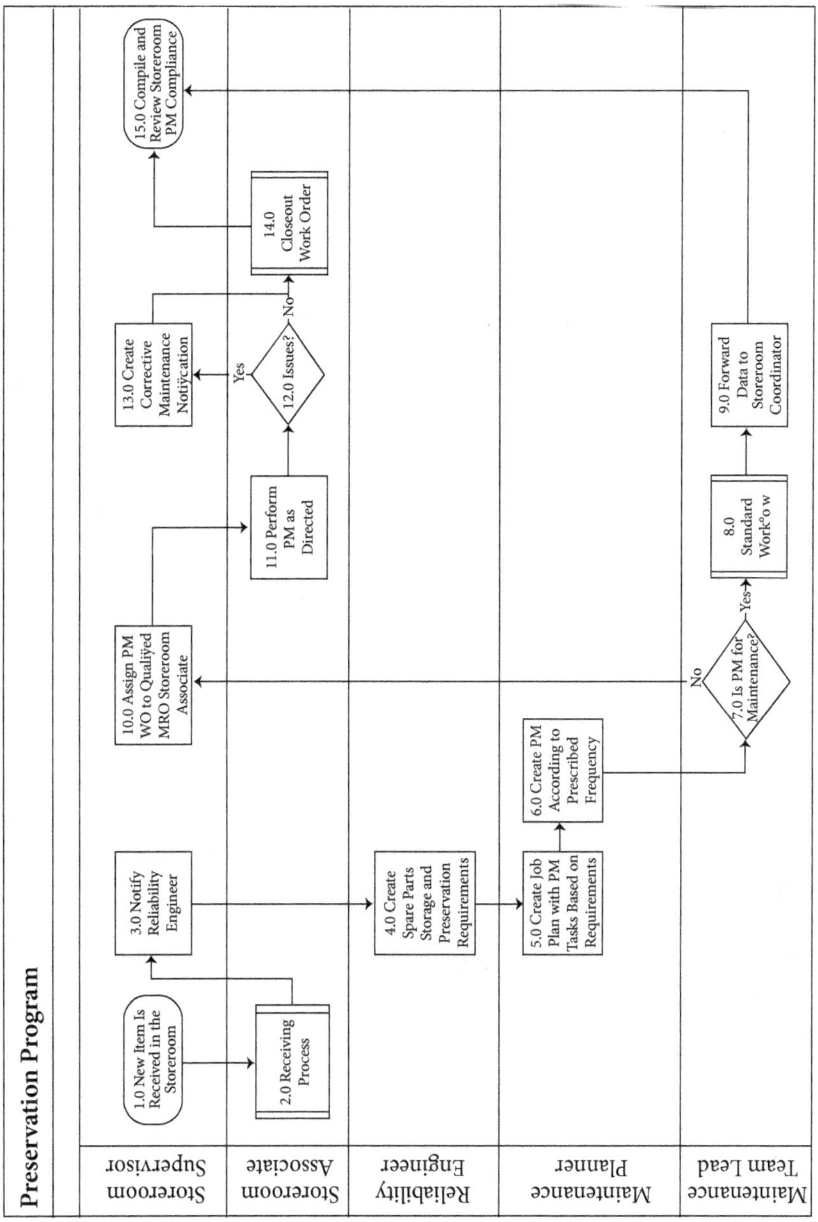

FIGURE 6-4 Preservation Program workflow.

First, consider that you don't have an infinite number of *types* of parts; it is very finite. You have motors, gearboxes, couplers, chains/sprockets, belts/sheaves, electronics, cylinders, pumps, etc. You can make a list today of the types of spare parts you have. I suggest that you do so. Ideally, the reliability engineer would have this list, and the responsibility to create a care strategy for each type of part would fall on that person.

I can make the job of developing the care scheme even easier. Let's assume that the following care environments are *not* suggested and should be avoided (to name a few):

- High temperatures, >85°F
- Inconsistent and wild temperature swings
- High humidity, >50% RH (relative humidity)
- Exposure to UV (ultraviolet) light
- Moderate or greater vibration
- Mishandling
- Propensity for damage by passing material handling equipment
- Dust buildup

If you look back at Figure 6-4, specifically at the reliability engineer's responsibility, you'll confirm essentially what I stated just a few paragraphs before. The person in this position should be an expert at determining the strategies to guard the inherent reliability of the components that are stocked in support of the operational assets in your facility. Where do these storage best practices come from?

Aside from exercising common sense by not storing parts in the detrimental environment just described, the OEMs of your components are your best sources for good information. They will undoubtedly confirm what you know to be commonsense practices and can provide other useful insights.

Nancy Regan's exceptional work, *The RCM Solution*, has a fitting thought on proactive maintenance tasks: "In order to schedule a proactive maintenance task in the context of RCM [reliability-centered maintenance], two criteria must be satisfied. The proactive task must be: technically appropriate [and] worth doing" (p.123).

Regan's book clearly is meant for asset reliability, but I don't believe it is too much of a stretch to apply that logic to preventive maintenance (a type of proactive maintenance according to Regan) tasks on components in the storeroom. Using the component OEM as a source, as suggested a few paragraphs above, would likely net PM tasks that are both technically appropriate and worth doing. Wouldn't you agree?

One of the more commonly known preventive care programs to execute on spare parts is to rotate motor shafts on a regular basis. This practice is usually meant to get the grease back up in the bearings, as grease is a liquid and will "slump" around the bottom one-third of the bearings if not exercised periodically. Also, the practice is recommended to avoid the false brinelling of the bearing races that would result from the weight of the rotor pressing down on the rolling elements and the races. How often should motor shafts be rotated? I recommend once per month. If you've stored your motors in a common location (see the sample storeroom layout in Figure 2-2) and have optimized your storeroom layout (the Intermediate-Effective zone), and have reduced your inventory count through Optimizing Stock Levels, this will be an easy 10-minute task. If, on the other hand, you have motors all over the place, this could take a while. Do you see how all of this is connected on the Storeroom Roadmap?

Figure 6-5 is from a facility that uses custom-made signage to indicate that the motor shafts have been rotated. For January–February, the number 1 on the placard will point at the 12 o'clock position. For March–April, the number 2 will point up, and so on.

FIGURE 6-5 Motor rotation signage.

This process located in the Advanced-Effective zone of the Storeroom Roadmap is meant to signal organizations to develop a process for determining if a component brought into inventory requires any special storage requirements to guarantee its survival while the part is in stock, and to ensure that it has its full life of service once it is installed for use. Secondarily, this process also is meant as a "check-and-see" note for the reliability engineer and others to confirm that whatever scheme you have in place to guard your component's inherent reliability is honestly working.

For those who have an interest, I'm providing a short list of care standards that I have found to be effective for various types of components. If you find or have other procedures that are effective at your location and have been proved to guard the lives of your components, please continue to use your own procedures. My input here is meant for you to have as a guide and possible discussion point around making your program better.

Table 6-4 is a short collection of motor PM tasks I recommend. It is likely that you are familiar with these.

TABLE 6-4 Motor PM Tasks

Motors
Protect shafts; again, protect the shafts.
Rotate shafts to protect against grease slump and false brinelling .
Plug injection points.
Guard keyways against incidental damage.
Keep the key stock that comes with the motor (affects balance).
Motors should not sit directly on the floor; use a pallet.
Keep motors in the box in which they came, if applicable.
Minimize the vibration of the surface that the motor sits on.
Protect the fan/brake cover.
Keep dry.

Table 6-5 is a list of gearbox PM tasks that I have seen to be very effective, and I recommend that you consider executing these if you do not already.

TABLE 6-5 Gearbox PM Tasks

Gearboxes
Protect shafts; again, protect the shafts.
Keep dry.
Plug injection points.
Guard keyways against incidental damage.
Keep the key stock that comes with the gearbox (affects balance).
Gearboxes should not sit directly on the floor; use a pallet.
Keep gearboxes in the box in which they came, if applicable.
Minimize the vibration of the surface that the gearbox sits on.
Watch and guard against any leaks.

Table 6-6 is a list of cylinder PM tasks and ideas that work well.

TABLE 6-6 Cylinder PM Tasks

Cylinders
Always store fully retracted, with large cylinders standing up.
Store indoors in a clean, dry space that has minimal temperature changes.
Protect the exposed piston rod, steel, and thread with rust inhibitor oil or grease.
Keep rod threads protected by either storing with preinstalled attachments or covering with protective sleeves.
Apply lubrication grease to any eye/clevis bushing or bearings—especially if they are steel.
Plug ports with metal plugs.
If possible, fill hydraulic cylinders with oil on the rod side only before plugging.
During the storage period, extend and retract the hydraulic cylinder several centimeters/inches (at least once per year), to prevent the seals from bonding. Depending on the results, you may have to take corrective measures.

Table 6-7 is a list of electronic PM tasks that should be common practice. Remember to keep your electronic components dry and cool.

TABLE 6-7 Electronics PM Tasks

Electronics
Store your electronic equipment in moisture vapor barrier (MVB) bags. Thicker is better, so buy the best you can afford.
Vacuum- and heat-seal the bags to remove as much air as possible.
Use properly sized desiccant bags to absorb any moisture, and include a moisture/humidity card that can detect any moisture. This will let you know if the storage bag has been compromised, meaning it is more likely the equipment in the bag may not work.
Keep your spares at a consistent and cool temperature—between 60 and 70°F.
Invest in the best moisture-free environment you can afford.
Eliminate all vibration; use vibration isolation pads.

Table 6-8 is a list of bearing PM tasks that should be the common approach to caring for these high-precision devices in your storeroom. They are highly precise devices even though they are treated roughly some of the time.

TABLE 6-8 Bearing PM Tasks

Bearings
Store bearings flat, in a vibration-free, dry area at a cool, steady temperature.
Control and limit the relative humidity as follows: 75% at 68°F 60% at 72°F 50% at 77°F
Keep bearings in their original unopened packages until immediately prior to mounting to prevent the entry of contaminants and corrosion.
Bearings that are not stored in their original packaging should be well protected.
Lubricant deteriorates over time as a result of aging, condensation, and separation of the oil and thickener. Therefore, sealed bearings should not be stored for more than three years.
Do not store bearings directly on the floor; use a pallet.

Figure 6-6 is from a facility that hangs its belts from pegs but in UV-protective bags. The bags are hung from the pegs, not the belts themselves. This serves two purposes: It eliminates the damage caused by UV light and gets away from the unsightly storage and damage-causing practice of hanging belts from hooks.

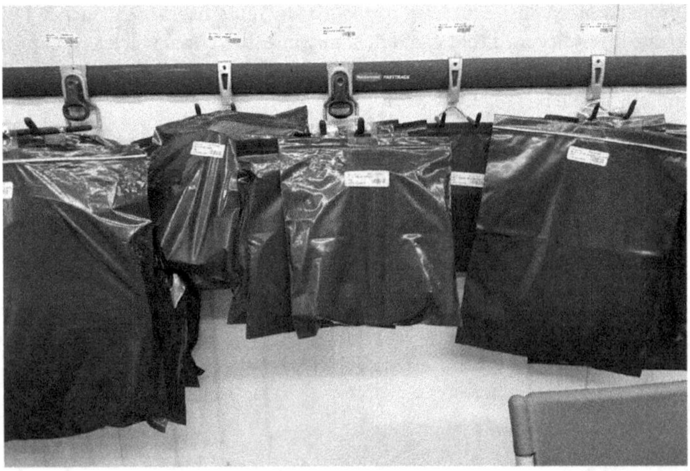

FIGURE 6-6 UV-protective bags.

Figure 6-7 is an image of damage caused by hanging a heavy belt on a peg for a long time. This downward weight and pressure create an indentation on the belt, which is a stress riser. Long, heavy belts are best stored lying flat. If you lay your belts flat, they are going to take up a lot of space. (See Optimized Stock Levels to reduce the number of belts and Optimize Physical Layout to have space for these belts.)

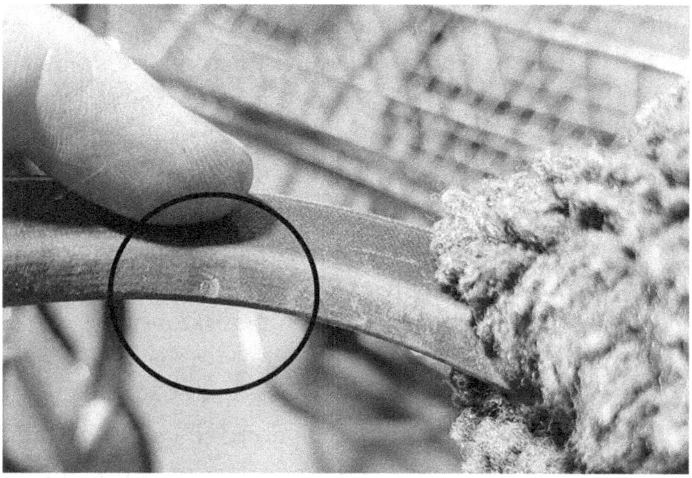

FIGURE 6-7 Stress riser in belt.

Table 6-9 is a potpourri of additional life-preserving steps I've seen other organizations use with great success.

TABLE 6-9 Additional PM tasks

Additional Thoughts on PM Tasks
Use a FIFO (first in, first out) methodology.
VFDs (variable frequency drives) might need attention to keep capacitors charged.
PLC (programmable logic controller) programs should be PM'd monthly and matched against a master version that is stored by IT.
Absolutely nothing should sit directly on the floor.
Use a vibration pen to check the vibration level of your shelving.
Nothing should be in a basement or in a damp environment or a dusty environment.
Avoid storing anything outside, especially rotating equipment.

A few pages ago, I wrote two short sentences that could have gone unnoticed. I want to restate those here because they are instrumental in delivering a sound PM program on spare parts in a sparsely staffed storeroom. Here is what I wrote: "The overall program will have activities to be conducted by *maintenance personnel* and those meant to be completed by storeroom personnel." Also, I wrote, "For example, highly technical tasks on variable-speed drives might specify that *a particular skilled trade is needed* to complete the PM."

You have seen this effort to enlist the skilled trades to augment the storeroom staffing before. It was part of the Cycle Counting process in the Intermediate-Effective zone. If you have a storeroom staff of one or even two, it might be possible, and even a good idea, for you to work with the maintenance department leadership to have a technician from time to time, as the PM schedule requires, perform the necessary steps to preserve the very spare parts that the technician might someday use. The technician's work would be covered by a work order (of course) and would be tracked and reported like any other preventive maintenance work order.

As I've done in the past in this book, I can make a strawman argument against this approach. How does one person assigned to the storeroom have the time to accomplish all the work that has to be done in the storeroom? How can that person do an exceptional job on everything? Look back at Figure 3-3, at the bottom series of connected workflows. Imagine one person doing all of that. The Preservation Program is just one of the workflow activities in that complex figure.

This was the last element of this zone, with only one more zone to go. The Advanced-Efficient zone is the end of the Storeroom Roadmap, but hopefully not the end of your journey.

SUMMARY

This was your first entry into the advanced zones, and as it turns out, there are not many elements in the Advanced-Effective zone. That makes sense if you think about it. All the other zones across the *effective row* constitute the lion's share of what is required to be truly effective at storeroom operations. The two elements in this zone are just the final push to get your program to the highest level.

We began this zone with the Inventory Turn Ratio KPI, or metric (this depends on how you have it set up). This is probably the most underused and misunderstood metric you will encounter in the storeroom. I tried to explain it using the metaphor of a store that sells nothing but bread. That explanation seems to work in my storeroom classes.

This is a very important metric to understand and, more importantly, to put into use. This is the metric that tells you that you are stocking about the right items in about the right quantity. You want your inventory to turn and replenish like a store that sells nothing but bread. This factor also aids in moving inventory before it exceeds any storage expiration date. Some studies have shown that the best companies can still lose about 2–3% of their stocked items due to exceeding a "sell-by" date or being damaged due to excessively long storage times.

The Inventory Turn Ratio was the last of the Big 3 KPIs that are collectively listed as KPIs on the list of 32 storeroom processes and practices. I strongly encourage you to track these three. Track more if you want to, but definitely the three that are documented in this text.

This zone ended with the most consequential element on the entire Storeroom Roadmap. The Preservation Program is the preventive care program for the items stocked in the storeroom. Not every item in the storeroom needs a PM program associated with it. I tell my clients that a block of aluminum probably doesn't need to be PM'd, but everything else does.

The reliability engineer is generally tasked with creating the scheme to protect components while in stock. An effort is made to guard the inherent reliability of the component. A part might be damaged in transit to your location and get compromised when it leaves your care, but while it is in the storeroom, it will receive the greatest protection. Does it make any sense to do otherwise?

WORKBOOK

The workbook for Chapter 6 is understandably shorter than those of the other chapters. There are only two elements in the Advanced-Effective zone. However, one element is a critical and often misinterpreted metric, and the other is the most consequential process on the map.

Here is the assignment for Chapter 6. Fill in Table 6-10 with a Y for yes or an N for no.

TABLE 6-10 Answer Y or N

	Y or N	**N/A**
Do you have formal process guides created for the following?		
Preservation Program		
Do you have the best preservation program possible for the following?		
Bearings		
Drive belts		
Motors		
Gearboxes		
Electronics		
Cylinders		
Does your storeroom look like you care about how things are stored?		
Do you measure inventory turns?		
Are your turn ratios around 1–2?		

Table 6-11 lists some short videos that cover the elements of this zone. Take time to view these videos, and ask your leadership team to create ways to use this medium of learning as you expand the knowledge and capabilities of your organization.

TABLE 6-11 Topic Links

Title	**Maintenance Innovators, Inc., Website Link**
Preservation Program	https://maintenanceinnovators.com/aiovg_videos/preservation-program/
Inherent Reliability of Spare Parts	https://maintenanceinnovators.com/aiovg_videos/the-inherent-reliability-of-spare-parts/
Repair Time . . . Spare Parts	https://maintenanceinnovators.com/aiovg_videos/repair-timespare-parts/
	YouTube Dr. John's Maintenance Minute Link
Preservation Program	https://www.youtube.com/watch?v=qE8SdtyqCS8
Inherent Reliability of Spare Parts	https://www.youtube.com/watch?v=Mb6Kfqwezto
Repair Time . . . Spare Parts	https://www.youtube.com/watch?v=igWNxOHgz6Q&t=15s

CHAPTER 7

ADVANCED-EFFICIENT

> Other leaders used their intelligence as a tool rather than a weapon. They applied their intelligence to amplify the smarts and capability of the people around them. People got smarter and better in their presence. Ideas grew, challenges were surmounted, and hard problems were solved.
>
> —Liz Wiseman

The end of the road. Well, not the "end" end. You'll attend a seminar or a conference, watch a video, or read another book that will inspire you to continue working to further develop your world-class storeroom. That's what continuous improvement is, to an extent. This zone is just where the paved road ends. Adventure and great things lie beyond.

You will find that the odd collection of organizational elements, practices, and a process included in this final zone are truly meant to make your storeroom more efficient. I can guarantee that absolutely every element in this zone is already in practice at your location, but this is the time in which you hone it and make it the best that it can be.

Figure 7-1 is the layout for this portion of the Storeroom Roadmap, and Table 7-1 is the associated legend.

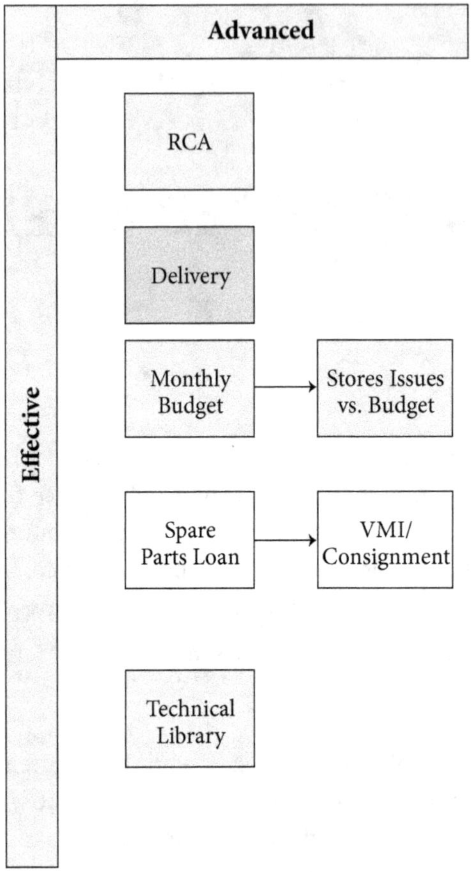

FIGURE 7-1 Advanced-Efficient zone.

TABLE 7-1 Legend for Advanced-Efficient Zone

Element	**Activity**
RCA—Root Cause Analysis	Organizational
Monthly Budget	Organizational
Stores Issues vs. Budget	Organizational
Delivery	Process
Spare Parts Loan	Practice
VMI/Consignment	Practice

We begin our final roadmap discussion with an element that will likely need continuous effort for improvement and one that you hope to never use.

RCA—ROOT CAUSE ANALYSIS

A colleague put it best when he said this to his class: "World-class companies are world-class problem solvers." I think this is spot-on and have used this quote many times. Most of you reading this book are working at world-class companies, but you might not be world-class problem solvers. Not yet.

No one reading this book can go through a single day without addressing or being asked to help solve a problem. You and I solve problems for a living. We know that we are going to spend every day solving problems, so why not get exceptionally good at it?

I challenge my students to become the Van Helsing of problem solvers. Van Helsing is the legendary fictional monster hunter sometimes referred to as the "Vampire Slayer." We are going to hunt down problems in their infancy and kill them while they are still small. Drive a stake right through their hearts, I tell my classes. Why? Because world-class companies are world-class problem solvers. And your company is a world-class company.

In my first book, *The Reliability Excellence Workbook: From Ideas to Action*, I explained the importance of RCA and its companion, problem-solving, and gave examples of their use and how to make each better. I started with the definition of each, and I think that is important here as well. These *street definitions* are from that publication (Ross, p. 155):

> **Root cause analysis (RCA).** The practice of analyzing data and information to arrive at the primary cause(s) for an unwanted event or trend.
>
> **Problem-solving.** The act of defining a problem and its consequences and creating at least one, but probably several, options to correct the issue.

It doesn't matter what RCA or problem-solving tool you and your organization use. Just become really good, and really effective, at their use.

I was once asked how you know you've arrived at *the* root cause. I said, "Because it doesn't happen again . . . for a while." I had to throw in that last part because a motor will fail again, a switch will fail again, etc.

As a student in maintenance and reliability myself, I've adopted another approach that I learned from a student: "You know you've found the root cause when you can turn it on and off, and the problem comes back and disappears." I think that might be a learning from Six Sigma.

I have an amusing story around this idea of "never seeing it again," or at least, of "it being the last one."

Years ago, I was in the market to buy a house in the Kansas City area. My real estate agent gave me a list of houses that matched my search criteria and asked me to pare it down to 10. So I set out to look at the houses and trim my list to a manageable 10.

One house was in a suburb of Kansas City that I was unfamiliar with. I drove to the little town and stopped at a convenience store to ask for directions. The 17-year-old girl at the register helped me with directions: "Turn left out of the parking lot, go to the last traffic light, and turn right. About a mile up the road on the right is a neighborhood. The street you are looking for is in that neighborhood."

I thanked her, left the parking lot, and turned left. I must have driven about 200 yards down the road; then I turned around and returned to the convenience store. Here is the next exchange I had with the young lady:

Her: "Did you have trouble finding the street?"

Me: "I just wanted to check the directions. You told me to turn left out of the parking lot, go to the last traffic light, and turn right. And about a mile up the road on the right is a neighborhood. The street I am looking for is in that neighborhood. Is this correct?"

Her: "Yes."

Me: "Just out of curiosity, how do I know when it's the *last* traffic light?"

Her: "Because there's not another one after it."

How do you know you found the root cause? There's not another one after it. And do you know something? She was right. There wasn't another traffic light after that one. In case you were wondering how this story ended, I did find the house, right where she said it was.

You aren't looking for an address in the storeroom. What you are looking for is quite a bit more elusive and subtle. There are many nuances to problem-solving and RCA in a storeroom that can't be solved by merely phoning it in, as is often the attitude toward completing RCA reports in the maintenance department.

That was, admittedly, a little harsh, but I've seen dozens, maybe hundreds, of half-effort RCA events with impotent after-action task lists that never seem to get accomplished. Is this you? Be honest.

For the definition of "root cause," I specifically mentioned that the tool is used when something is *trending* in the wrong direction. We don't need to wait for something to be broken in the storeroom to act. We are looking for activities that are *trending* in the wrong direction.

Table 7-2 is a short list of values that might trigger an RCA action if they were to be trending in the wrong direction.

TABLE 7-2 Storeroom RCA Trigger Issues

Increase in stockouts
Increase in kits missing items, incomplete kits
Increase in items removed from the storeroom without documentation
Increase in inaccurate cycle counts
Failing to reach a critical metric or KPI for two or more months
Increase in incomplete New Item Setup forms
Increase in vendors'/suppliers' late or partial deliveries
Increase in damaged items received

It is important that you, along with the Stores Stock Committee, establish trigger points for the storeroom's performance and that the SSC is engaged in helping to solve problems and find root causes.

Table 7-3 contains a list of RCA tools I've used or seen used in the years I've been engaged in RCA and problem-solving. This is not an exhaustive list, but a list of what you might encounter or tools you may currently use.

TABLE 7-3 RCA Tools

RCA Methods	**Data Analysis Tools**
5 Whys	Histogram
Fishbone	Scatter diagram
Cause and effect	Pareto chart
A3	Day-by-the-hour chart
Apollo©	Run chart
TaprooT©	
DMAIC	

It doesn't matter what tool you use; get good at using it. One common element within all the tools mentioned in Table 7-3 is the need to create a solid problem statement.

I have worked with a lot of companies to help them solve problems, or more proactively, to help them create a culture of problem solvers and hone some problem-solving skills with the tool of their choosing. I do this through a workshop, what I like to call a "live fire" workshop where we work together on an actual issue.

It is not uncommon for me to hear that the problem that I am there to help the companies solve is along the lines of something on this list (and this is a truncated list):

- The storeroom never has anything.
- That pump never works.
- Production runs everything into the ground.
- This place doesn't care about safety.
- People are stealing stuff out of the storeroom.
- Maintenance never fixes anything.

These statements collectively have the advantage of being the worst problem statements ever. The first thing you are going to do is get the problem statements right. From this day forward, never accept "That pump never works" as a problem statement. Really? That pump never works? Do you mean that it's not working now?

Or my personal favorite, "The storeroom never has anything." I heard this from a mechanic at a facility while I was performing an assessment and audit of the maintenance organization. The storeroom carried 47,000 line items with a combined value of $75 million. It was and remains the largest storeroom I've ever seen.

The backstory, and I wish I were making this up, is that the mechanic told me he had gone to the storeroom two years before to get a red, permanent, fine-tip marker. This was an item that the storeroom stocked. As it turns out, the storeroom was out of those pens that day. And since then, this mechanic was convinced that the storeroom never had anything.

A problem statement provides the foundation upon which to build the logical solution to the issue. You must become exceptional at defining the problem. Here are the five fundamental questions that must be answered in the problem statement:

1. Who discovered the issue (*who*)?

2. How big is the issue (*what*)?
3. Where is or was the issue (*where*)?
4. What time frame was the information collected in (*when*)?
5. The issue resulted in what, exactly (*resulting in*)?

As an example, Atlas Company is suffering from a negative trend of maintenance technicians not signing out parts when they take them from the storeroom. The storeroom leadership is likely to define the problem statement as:

> Maintenance technicians never sign out the parts they take.

That is a statement and a grievance I've heard a thousand times if I've heard it once. You're likely to have said it yourself a time or two.

Here is a better, well-defined problem statement that answers the questions listed earlier:

> The day shift clerk recorded eight incidents where items were removed from stores without recording the work order number or the name of the person who removed the items. This was recorded over a two-week period, and all the incidents happened on the midnight shift. This resulted in the absence of issuing documentation and the failure to replenish inventory on six items because the reorder point for those items was not flagged.

I hope you will agree that the second problem statement is clearer and fuller. And that the second statement gives us more insight into what truly happened.

Answer the five fundamental questions for a good problem statement using the first statement provided. You can't, can you?

Let's look at the same exercise using the second problem statement:

1. Who discovered the issue? *The day shift clerk.*
2. How big is the issue? *Eight incidents where items were removed from stores without recording the work order number or the name of the person who removed the items.*
3. Where is or was the issue? *In the storeroom.*
4. What time frame was the information collected? *This was recorded over a two-week period, and all the incidents happened on the midnight shift.*

5. The issue resulted in what, exactly? *The absence of issuing documentation and the failure to replenish inventory on six items because the reorder point was not flagged.*

The question about the problem's *result* is significant for the sole purpose of understanding how much we care about the problem. Removing items from the storeroom and not recording a goods movement will eventually result in a missed reorder point (or two). This could lead to a stockout or possible extended downtime and even greater potential for the company's revenue loss. This contributes to the perception that the storeroom never has anything.

After all, it doesn't even have red, permanent, fine-point markers when it should.

I am borrowing the following exercise from *The Reliability Excellence Workbook: From Ideas to Action* (Ross, p. 163). See how you do in creating an exceptional problem statement.

Scenario:

The support team lead technician is primarily responsible for ensuring that the plant's compressed air is held at 80 PSI during all phases of plant production throughout the facility. The support team is responsible for the plant's two air compressors. During normal operations, one air compressor is running all out, and one is trimming over the top. The manufacturing facility operates 24/5, choosing not to run during the weekends and holidays. The support team ensures the air compressors are online two hours before startup on Monday mornings. Operations executes a staggered startup protocol. The lead technician noted that the two plant air compressors have failed to maintain a plant air pressure of at least 80 PSI on three separate occasions. These incidents have occurred over the last four weeks, and each time was during plant startup on a Monday morning. Because of this issue, there has been a repeated delayed start of 20% of the plant's machinery, resulting in a $20,000 loss of potential revenue.

Answer these five questions:

1. Who discovered the issue (*who*):

____

2. How big is the issue (*what*):

 ____

3. Where was the issue (*where*):

 ____

4. What time frame was information collected (*when*):

 ____

5. The issue resulted in what, exactly (*resulting in*):

 ____

Here is what I came up with:

> The support team lead technician noted that the two plant air compressors have failed to maintain a plant air pressure of at least 80 PSI on three separate occasions. These incidents have occurred over the last four weeks, and each time was during plant startup on a Monday morning. This issue has resulted in the delayed start of 20% of the plant's machinery, resulting in a $20,000 loss of potential revenue.

Don't ever write a poor problem statement again or accept one. You now know better.

Even before you draft a problem statement, the first issue you will encounter is forming a team of engaged and enthusiastic associates. Let's face it. Not many folks are fired up about being part of an investigative board. Imagine the low energy you'll likely encounter when you announce that the trend you are trying to correct is "too many parts being returned to stores with the job kits." You can almost imagine people running from your first team meeting.

How do you get people excited about being on a problem-solving team and get them engaged in root cause analysis?

I don't know if this is the answer, but I want to share it with you anyway. It might work, and then again, it might not.

As I have previously mentioned, I was an aircraft crash investigator in the Air Force for many years. That isn't a permanent position. An aircraft incident investigation team comprises several officers, and I was one of just a few trained in the forensics and mechanics of aircraft incident investigations. The training is at an elite level, and only the crème de la crème are invited to attend. The competition for aircraft maintenance officers to get into the course is tough.

An investigative team is called a "board," and the board comes together when an incident occurs. All the officers have regular assignments and are at Air Force bases all over the world. They come together to investigate an incident and find the aggravating circumstances that must be addressed before another similar incident happens.

As I mentioned, the school I had to attend to be qualified and certified was extremely hard to get into. I saw it as the highest level of competence in my career field. I was an aircraft maintenance officer, and I thought that being able to piece together the facts after an incident must be the absolute highest form of professional discipline that there is.

Tokutaro Suzuki's edited work, *TPM in Process Industries,* explains how being trained in abnormalities can make you extraordinary: "Skill is the ability to do one's job, to apply knowledge and experience correctly and reflexively in all kinds of events over an extended period. Systematically accumulating training, experience, and information enables a person to exercise good judgment and act appropriately. *The more swiftly a person can deal with an abnormality, the higher the skill level*" (p. 262).

My point about being on the tip of the sword as far as being at the height of professionalism comes from my understanding of Suzuki's words. There is no greater abnormality for aircraft operation than a crash. Investigating it swiftly and completely takes the highest level of skill.

If that is true, and I believe it to be, why doesn't that translate to the civilian world and all the abnormalities occurring there that should be addressed? That is where we can start to build our engaged, elite, and enthusiastic vampire slayers.

Let's agree in principle that there is nothing exciting about being on a safety investigation where someone broke a bone or cut a hand. No one is rushing to be on that team. Likewise, you will never find a volunteer to help investigate why the storeroom door was left open 17 times last month. Face it. We don't look at incidents and accidents the same way that the Air Force or other folks do who know world-class companies are world-class problem solvers.

I suggest you consider forming a professional group of problem solvers in your facility. Like the Air Force did. Air Force training was exclusive and by invitation only. You (I) had to be selected out of a pool of hundreds. It was a feather in my cap to be selected and trained. It was a big deal. Let's make the investigative teams that solve problems and decipher root causes an elite force in your companies.

Everyone you will ever need to have on a storeroom Root Cause Analysis team is represented by a decision-maker on the Stores Stock Committee (remember those folks). What if you formed a team of stakeholders who have an interest in the storeroom's ability to quickly identify potential issues and snuff them out while they are small? This team would also attack larger problems or issues if they happened to crop up. This team would be by invitation only and would receive only the highest RCA and problem-solving training, including a dip into critical thinking. Team members would be certified problem killers. Do you want to be on this team? You'll need to wait to be invited, and even then, someone must retire before there is an opening.

You should make membership of a problem-solving committee a "must" and create a true level of "professionals only." I never wanted to hear of an aircraft being lost or of an incident involving an airplane or personnel. But if it did happen, I wanted to be called for the team. Why? Because I was a trained professional ready to hunt monsters.

We investigate issues and incidents, not to find blame with someone but to keep the incident from happening again.

I promise if you approach every negative trend or bad experience from the perspective of keeping it from happening again, only engaging your best in the solution, and making sure your people know that there is prestige in being picked for the team . . . you'll have investigative energy and very successful results.

There are two remaining organizational elements in this zone, I'd like to discuss those next.

MONTHLY BUDGET

I did not learn about the positive impact a high-performing storeroom could have on a maintenance budget until I was well into my years of professional maintenance and reliability. I spent the first decade in the Air Force, where, quite honestly, I had no input on the budget, let alone what we stocked. Fly and fight. That was all I was concerned about. I didn't even have a storeroom during my first years as a civilian and as a plant engineer/maintenance manager. It was each man for himself. One day, a maintenance consultant sat down and explained the entire budgeting concept to me, and the scales were lifted from my eyes.

The *SMRP Best Practices* publication indicates that up to 50% of a maintenance budget can be spent on parts and materials. The category is "Maintenance Material Cost," and the definition is "The total cost incurred for materials, supplies and consumables needed to repair and maintain plant and facility assets for a specified time period, expressed as a percentage of the total maintenance cost for the period" (p. 281). "Best-in-Class Target Value 50%" (p. 283).

That definition is further detailed as:

> The cost of all maintenance, repair, and operating material (MRO) used during the specified time period. Includes stocked MRO inventory usage, outside purchased materials, supplies, consumables and the costs to repair spare components. Also includes materials used for capital expenditures directly related to end-of-life machinery replacement so that excessive replacement versus proper maintenance is not masked. Does not include material used for capital expenditures for plant expansions or improvements. (p. 281)

I have often looked at that definition and the best-in-class target value of 50%. Most of the time in disbelief. It was, and still is, almost impossible for me to comprehend that a maintenance organization with an annual budget of $2 million could spend up to $1 million on repair materials alone and be classified as best in class for this metric. That is an important metric to know, and figuring out how to work with this knowledge is very important.

Interestingly, Dr. Blache, Director of the Reliability & Maintainability Center at the University of Tennessee, examined this metric in a study of 200 facilities. Dr. Blache reports, "Top-quartile facilities have only about 20% of their total maintenance cost tied up in parts. That figure is 50% for bottom-quartile facilities" (p. 6).

So is 50% truly a best-in-class target? Whom to believe?

Very effective maintenance organizations will look for opportunities to trim their maintenance expenses by 3% per year. It is important to make and understand the distinction between trimming maintenance expenses and cost-cutting. Cost-cutting is exactly what it sounds like. Trimming expenses requires a purposeful and pragmatic review of the expense categories and determining where there is an opportunity to improve. Value for money.

Table 7-4 shows the budget category breakout for a typical maintenance department. The table also reflects the *direction* at which these categories should be aimed from year to year. *Increasing* indicates a category that a maintenance organization should try to increase in spending, and *decreasing* indicates that a maintenance organization should try to decrease spending. I've included a column for rationale as well.

TABLE 7-4 Maintenance Budget Categories

Budget Category	Increasing or Decreasing	Rationale
Maintenance Labor	Increasing	Hourly associates should get annual pay raises
Overhead	Increasing	Salaried individuals should get annual pay raises
Maintenance Training	Increasing	Training budgets should increase to 4% of annual compensation
Parts and Material	Decreasing	Parts and materials should be less than 50% of the maintenance budget
Overtime	Decreasing	Overtime should be less than 5%
Contractor Costs	Decreasing	Contractor cost should be reduced as much as possible
Equipment Rental and Lease	Steady	Worse case, hold this steady

Table 7-5 is a popular exercise from the American Reliability & Maintenance Institute's (ARMI) *Certified Storeroom Manager/Technician* program (Day 3, p. 50). I use this example in all my storeroom classes and storeroom consultations. This figure shows the typical maintenance budget categories. See if you can calculate what a 3% decrease in maintenance expense would be. Using that information, draft what you think the following year's budget goals might be. Remember to use the information (increase/decrease) shown in Table 7-4 as a guide.

TABLE 7-5 Sample Maintenance Budget

Category	Current Year's Budget Value	Next Year's Budget Value
Maintenance Labor	$825,000	
Overhead	$175,500	
Maintenance Training	$95,000	
Parts and Material	$2,056,000	
Overtime	$17,500	
Contractor Costs	$230,000	
Equipment Rental and Lease	$45,000	
Total	$3,444,000	

I calculate that 3% of this year's maintenance budget of $3,444,000 would be $103,320. That would make next year's budget goal $3,340,680 ($3,444,000 – $103,320). Table 7-6 is my result from this exercise. Since this is a book on storeroom processes and practices, note specifically the 10% planned reduction in parts and materials spending.

TABLE 7-6 Sample Maintenance Budget

Category	Current Year's Budget Value	Next Year's Budget Value
Maintenance Labor	$825,000	Increase by 5% to $866,250
Overhead	$175,500	Increase by 5% to $183,750
Maintenance Training	$95,000	Increase by 15% to $109,250
Parts and Material	$2,056,000	Decrease by 10% to $1,850,400
Overtime	$17,500	Increase by 10% to $19,250
Contractor Costs	$230,000	Remain at $230,000
Equipment Rental and Lease	$45,000	Remain at $45,000
Surplus		$36,780
Total	$3,444,000	$3,340,680

Let me give you some insight into how *not* to reduce spending on parts and materials. Buy cheaper parts or reuse old parts. That is not how you maintain the inherent reliability of an asset. Some of you might feel inclined to highlight that last sentence and show it to your boss. Please do.

As a professional, your challenge is to figure out how to reduce spending on parts and materials. The answer lies outside of the storeroom.

The first and most successful way to reduce maintenance spending on parts and materials is to use fewer parts and materials. That sounded less cheeky in my mind. If your organization could use fewer parts and fewer maintenance materials, it would make sense that this category in the maintenance budget would be reduced. But how do you do that? One simple word. *Training.*

Palmer makes this point clear in his suggestion that organizations move to improve maintenance: "In all the discussion about how the storeroom should improve in a proper manner, it would be remiss not to mention the maintenance force's responsibility. Improving maintenance to create a more reliable plant with fewer breakdowns gives the plant more leisure to work on proactive work orders and also requires fewer parts in the first place" (p. 345).

I think you will agree with me that as a profession, we are losing the skill of troubleshooting. I don't know if that is a result of the newer generation being less patient to learn the art of troubleshooting, or the time crunch in manufacturing now that leaves little time to perform it. The result is that we are growing a generation of parts replacers. The heck with troubleshooting; replace these five parts so we can get back up and running.

Aside from the obvious cost associated with that type of operation, there is no real ability to perform analysis or RCA on any failure because you aren't certain which part or combination of parts fixed the issue. Also, the planner cannot take feedback from that maintenance job and craft a better-planned job for the next time it's needed.

What is needed and required is better troubleshooting skills. I've mentioned to several clients and classes that we train our maintenance technicians in the proper way to perform maintenance, but there is little if any training to hone fault-finding abilities. If everything worked as it should, I could be a master mechanic or electrician.

Do you remember the quote from Suzuki during the discussion on RCA? "*The more swiftly a person can deal with an abnormality, the higher the skill level.*" I challenge you to create a training environment that includes the art of troubleshooting. Start to track the number of parts used to fix issues and concerns for planned and unplanned work. Track this as a metric to demonstrate whether the training is working or not. I think you'll be pleased to see that it works.

Another opportunity to lower the cost of parts and materials for maintenance use is to purchase the same part for a better price. Shop your needs around. I think suppliers and vendors are more than accommodating and will work with you to get you a better price on the same components you are buying. Be careful not to jump around too many times. A vendor will always be willing to work with you on pricing, but the vendor might need a bigger piece of the pie. This isn't a storeroom function as much as a purchasing or procurement department function. It might be a good corporate initiative if there are gains to be made across the entire company.

Another chance to decrease the cost of parts and materials is to reduce emergency purchases. Sometimes it costs more to fly a part to your facility than the part cost to begin with. I mentioned during the discussion on the element of Emergency Procurement (Chapter 4) that anecdotally, <1% seems to be the best-in-class target value for this process. This is another great area to improve on. Not just to save money on parts and materials, but improving this metric indicates that you are starting to get your act together.

Look at Table 7-6 one more time. I want you to fully understand this budgeting scheme so that you can make a compelling case to your team and your organization. Notice that the only category that is recommended for a decrease is the parts and materials category. From that 10% suggested reduction in parts and material, your organization can afford a pay raise for the hourly associates (labor), a pay raise for the salaried individuals (overhead), an increase in training funding, and a little more overtime funding.

That is the impact of controlling this vital cost of doing business. No one is positioned better to help maintenance maintain a budget or improve a budget than the professionals in the storeroom.

Do not forget that.

STORES ISSUES VS. BUDGET

As an organizational element, I believe this is the most underappreciated feature on the entire Storeroom Roadmap. I was completely ignorant of the knowledge right at my fingertips or the value that planning and scheduling can provide.

Make careful note of the placement of this element relative to the Monthly Budget element. A maintenance organization that can control its budget is an organization that will have few surprises.

I've mentioned a few times that this isn't a book on maintenance practices, but I do invoke some information from my other books, and those of other authors, to put things into perspective. For example, companies that are good at planning and scheduling are exceptionally good at preventive and predictive maintenance. If you're good at PM/PdM and planning and scheduling, it follows that you'll have a very good storeroom. These three competences seem to exist together. It almost never fails.

If I can predict when I need a part and control access to equipment, I can ensure part availability when and where I need it. Furthermore, I can estimate, with great accuracy, how much a maintenance job is going to cost. Part of the cost of a maintenance job is the cost of parts and materials.

Put all this together, and this is what you will understand. Your maintenance scheduler can tell you, within a range, how much money the maintenance department is *scheduled* to spend next week. This is based on the jobs that are scheduled to be completed. If your maintenance department has $196,000 to spend each week, inclusive of all costs, your team can determine if next week's schedule will be within budget. A maintenance organization cannot do this high-level forecasting if the department can't correctly sign a part out of the storeroom (as an example). I'm mixing metaphors here, but I hope you get my point. Repeat this 52 times a year and you end the year within budget.

The storeroom contributes information on the cost of the part for the overall maintenance cost equation. Where is the cost of a part listed? The Item Master Data. Remember Data Scrubbing all those zones ago?

When I was a new civilian maintenance manager, I would absolutely cringe at the thought of going to a monthly budget meeting with the plant manager and the leadership staff. Our plant controller would not hesitate to point out that maintenance was over budget. Sometimes we were within budget for the month. More likely than not, we were over budget. If you recall, I was in the Air Force at the start of my professional life, and budgeting and budgets were not topics I was involved with. Let's assume that the Air Force is always over budget. Once I created my first storeroom and hired my first storeroom lead (it was a union plant), I began to understand what the storeroom could do for me and my maintenance budget.

My storeroom lead was a lady named Kathy. Kathy was not a maintenance person; she was an operator who got the bid for the job because her clock number was 3, and the other two, more senior folks, had died. As I said, this was a union operation, and that's how it worked back then.

Kathy would walk over to my office from time to time to tell me that a technician had just removed *X* dollars of inventory and she thought I ought to know that.

Without even knowing that Kathy was feeding me "intel," I became a little more comfortable at the monthly budget meetings. If I was over budget, I would know why. One of the reasons I was dancing around the budget limit, to begin with, was because of spare parts and materials. As you have already learned, it can be up to 50% of your budget. For me, that was an uncomfortable level.

After a bit more education, and proactive steps with planning and scheduling, I was more confident of my budgeting ability, and the results were positive.

The storeroom can aid the maintenance department in maintaining a budget by keeping tabs, reporting on storeroom issues, and keeping costs accurate and up to date for cost estimates. Also, by alerting the maintenance leadership when high-value inventory items are issued out of the storeroom, the organization can note this and take follow-up actions if necessary. This is a vastly untapped opportunity.

I want to say one more thing before we completely leave the budget topic. You can't expect your team, either maintenance or the storeroom, to help maintain a budget if they don't know what the budget is. I strongly encourage the maintenance leaders to inform their entire organization of the budget, the categories of the budget, and the budget goals. Let the members of your team know that you are going to be tracking expenses and gathering and analyzing data to guide the budget "ship." Ask for their help. You know the people on your team better than I do, and I'm sure you won't be surprised that many will help you once they understand it.

One other organizational element in this zone is the Technical Library, which I already addressed when discussing the Intermediate-Efficient zone's non-materials storage category. This element is in the advanced zone, but I thought I would keep continuity with the non-materials items.

With these final organizational elements, we end our discovery of these very simple yet impactful roadmap features. At the beginning of this book, when I provided some street definitions for the elements and zones you would encounter on this road trip, I shared my thoughts on organizational elements. These are operating features that are generally necessary for any functioning entity. Like every other segment of business, the storeroom requires sound management, leadership, and background workings to round out smooth operations.

If you look back at the elements I've labeled as organizational features, I believe you will agree with me that their implementation and improvement will help to round out a storeroom that is both effective and efficient.

We have three elements left to see on our adventure. One of them is the last tactical process we will encounter.

DELIVERY

The highest point of evolution for the Kitting process is the delivery of those kitted items. For a task as simple as delivering the mail, this tactical process is an element in the advanced zone for good reason. It takes a lot of brain power and willpower to consistently pull this off. Doc Palmer prepares readers to understand the complexity of staging: "Staging can become somewhat complicated to execute properly" (p. 354).

All items delivered by the storeroom must be planned, kitted, and issued to a valid work order and tracked against the work order's cost center coding for accountability.

This delivery process requires that the Stores Stock Committee establish the location for the transfer of item(s) ownership between the storeroom and the work executor (most often the maintenance department) and establish the restrictions around the delivery meant to keep personnel safe and property secure.

Key to the successful execution of the Delivery process is preapproved and sanctioned delivery locations. The preapproval is suggested by the parties involved in the giving and getting of the kits. The sanctioning is provided by the SSC as alluded to above. Your SSC should designate certain locations around your complex as approved and guarded kit relocation and storage locations. Much like the safety perimeter around electrical panels, the space set aside for the delivery of kits should be held as sacrosanct.

The risk associated with getting this simple transfer wrong is similar to an elevated risk associated with any transfer of ownership. In a typical storeroom service window exchange, the storeroom owns the part until the exchange of "work order for the part" takes place through the window opening. It is an "I have it, now you have it" arrangement.

For kitted and delivered items, the understanding of this exchange must be clear. When does the storeroom own the kit and therefore have the responsibility for its completeness and the part's reliability, and when do you own it and take that responsibility?

Hopefully, I've been successful in painting a picture that simply dropping off a pallet of expensive parts in a dark alley is not going to cut it. Not in an advanced storeroom.

Delivery locations are sometimes short-term storage locations. These spots might be the final staging points before executing a maintenance job. These sanctioned delivery locations should be covered and offer protection from incidental damage (e.g., a wayward fork truck) and inclement weather. Don't forget that a job kit contains valuable spare parts. Some of these parts might be high-priced and low-availability.

More often than not, industries that require that parts and job kits be bundled and then delivered by the storeroom personnel are industries that typically have large geographic footprints. Oil fields, refineries, chemical plants, universities, and large manufacturing operations often run a delivery service in or out of the storeroom.

In Chapter 2, I mentioned a well-running storeroom operation in South Texas when discussing the Receiving process in the Foundation-Effective zone. The storeroom manager was a disciplined leader and very process-driven. I wrote in that section that the lady in charge of this vast storeroom had an exceptionally effective receiving process. That storeroom had another enviable characteristic.

I was there for six months, and I was joined by a few colleagues on the project. I can't speak for them, but while I was there, I never saw anyone come to the service window to request a part. I asked the storeroom manager why I had never seen anyone at the window. She told me that someone would only need to come to the window if there were an emergency. "John," she continued, "we don't have emergencies at a chemical plant."

This chemical plant operation was practiced and proficient at preventive maintenance, job planning, scheduling, pick list generation, and kitting job kits. The storeroom had a fax machine that was only used by the 13 planners (you read that correctly—13) to send their pick list to the storeroom. There was an associate in the storeroom who had the primary responsibility of retrieving the fax and putting the kitting wheels into motion. (See "Staffing Roles and Responsibilities" in Chapter 2.)

The storeroom had at least two delivery drivers who spent the entire day shift making their designated rounds. They would drop off job kits and pick up empty containers or items that needed to be returned to stock or sent out to be repaired.

The secret to the timing of everything was the fact that the storeroom was instructed when to kit the jobs and that the storeroom held the job kits until the day before the job was to be executed. Then the driver would deliver the kit to the planner the day before the job. The planner in each of the 13 locations had a small, secure holding area.

The orchestration of this Kitting and Delivery process dance was beautiful to see.

It works great when it works great. It is a disaster if you aren't committed to it. That's why it is an advanced process. Don't try it if you don't believe in it or don't believe it will make you more efficient.

We have two elements left in this zone and on our Storeroom Roadmap. They are both strategic practices.

SPARE PARTS LOAN

This strategic practice is often the most controversial. And for good reason.

The Spare Parts Loan practice is necessary to ensure that your organization has a disciplined approach to loaning parts to sister facilities, and/or outside your organization. Often to competitors. That is the controversial part.

Before an item in inventory can be loaned to an outside organization, rigorous research must be conducted into the impact of that loan. This should involve using a risk matrix or another type of objective measuring tool. When the replacement item is received, it is scrutinized to certify that it meets your needed specifications. Outside organizations with credible requests for a loaned item will be required to provide plausible reasoning for requesting the loan and documentation indicating they have the item on order. It would not be unusual for a sister facility in your organization or even a competitor to buy its parts from the same supplier you buy your parts from. This makes sharing and loaning parts within the same industry an even better proposition for your organization to investigate and understand.

Figure 7-2 is a good example of how the Spare Parts Loan work practice should begin.

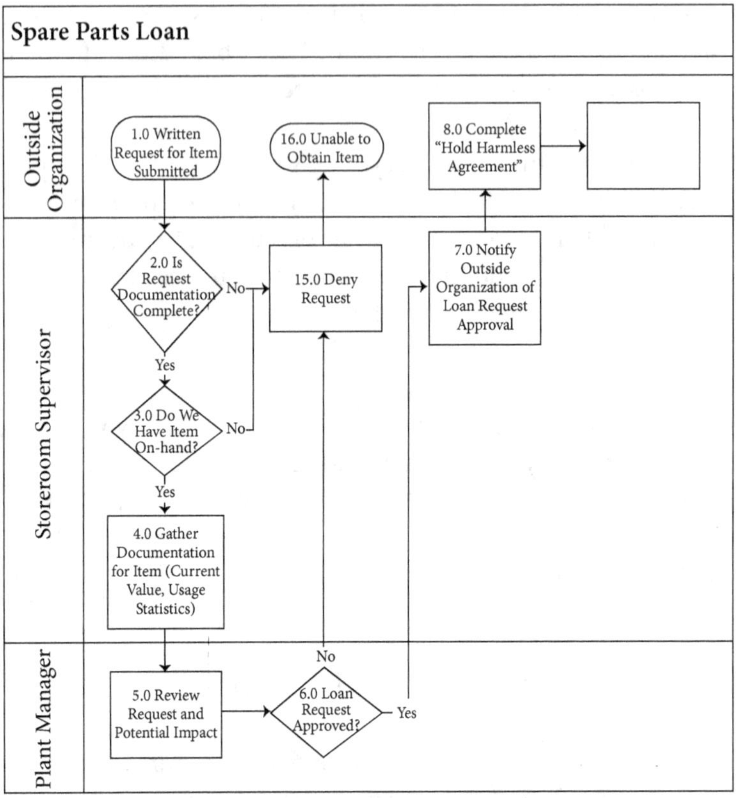

FIGURE 7-2 Spare Parts Loan workflow (partial).

I mentioned a few times that this was controversial because very few companies would knowingly help their competition. But I've seen this quite often in the liquefied natural gas industry. As it turns out, what's good for your industry is also good for your company. Someday it might be your facility looking for help.

"Loan" might be the wrong word in this case. There should not be an expectation that you are going to get back the very part that you loaned out. In fact, I would caution against it. You do not want a scenario where you loan a part to a neighbor, the neighbor uses that part until the neighbor's new part arrives, and then the neighbor gives you that (now) used part back. That would not be ideal.

Looking back at Figure 7-2, here are some elements we can agree on, whether "loan" is the right word or not:

- The request from the outside organization must be detailed and in writing.

- The loaning organization (yours) must have the part; it could be that you have a part coming in soon, and that is the part being considered for loan.
- The plant manager should evaluate the risk to your organization. I recommend using a risk matrix and for goodness' sake, never loan a part without the plant manager's approval.
- The requesting organization must have the part on order, and there must be a verified delivery date; this is used for your team to consider the amount of time you are likely to be at an elevated risk (no spare part on hand).
- When your organization receives the new part that was originally coming in for the requesting organization, make sure to thoroughly inspect it for fit and function.
- It may also be that there is no risk at all.

Including this as a practice on the roadmap is not to indicate that this is an activity that you should be involved in at least four times a year. Rather, it is a practice because it is hoped that you'd never be required to use it. There are enough questions about the supply chain and its sustainability to warrant every organization having a policy about loaning parts and requesting parts to be loaned. At some point in a long career, you will be impacted by this practice, as the giver or the receiver.

This practice is in the advanced zone because you shouldn't even be thinking about doing this unless you can execute the Foundation-Effective zoned elements first. This practice is as far as you can get on the map from the initial efforts it took to set up your storeroom. Don't loan out parts if you can't set up parts in your own storeroom.

Also, notice the subtle connection with the Parts Standardization element in the Intermediate-Efficient zone. Do you recall the definition of "proactive" from Tom Moriarty that I've shared a few times? Imagine if your organization, across perhaps many plants, had a proactive approach to parts standardization, and you didn't need to necessarily have the part in stock yourself to have access to it. Now your organization has an established and practiced procedure for loaning and borrowing parts. This gets very close to one of the highest categories of spare parts . . . the strategic spare.

In the context described above, a strategic spare is what you imagine it would be. The example I often share is of a company with many operating plants, and each of those plants uses a particular 100-hp motor. It is the same motor in each plant. Your company doesn't need to have a spare

100-hp motor in each plant, just a few positioned around the country as *strategic spares* to be called into service if necessary. The plants where the spares are stored will be responsible for *preserving* their inherent reliability and processing the motors for shipment as needed. The holding facilities would be resourced to handle this additional, corporate-mandated assignment for moving parts around the company. This is very high-level.

Our Storeroom Roadmap ends with the *strategic sourcing* practice of *VMI/Consignment.*

VMI/CONSIGNMENT

This practice aims to further improve the storeroom's efficiency by increasing the VMI and Consignment possibilities. At this point on the roadmap, your organization must certainly be utilizing these strategic sourcing methods. Vendor Managed Inventory and Consignment are meant to leverage purchases, vendor relationships, and customer needs.

This practice is typically owned by the procurement or purchasing department with consultation from the maintenance department as the primary end users. I strongly recommend that the Stores Stock Committee be engaged in the advancement of VMI and Consignment services. Through the SSC's understanding of the way in which these two methodologies work, they can better see the value. The finance folks are often the owners, since contracts are involved and terms of payment become serious business. More so than usual.

"VMI" refers to inventory on the shelf in the storeroom, but the vendor performs its upkeep, rotation, fulfillment of stock, and cycle counting. VMI stock typically consists of low-cost and mostly nonproduction consequential items.

Examples would include:

- Filters (not on critical assets)
- Belts (not on critical assets)
- Bench stock items (nuts, bolts, washers)
- Piping fittings
- Earplugs and other safety-stocked items

I don't recommend having critical spare parts, or any other part that is vital to a critical asset, stocked using VMI as the fulfillment methodology.

It is too easy to get overconfident in this situation and get caught in a crisis. Keep a firm *company* hand on critical spares and on components to critical equipment.

With VMI, your company owns the stock; the vendor manages it. VMI items will be on open purchase orders. There is no requirement for all VMI items to have a part number or a catalog number.

Bench-stocked VMI hardware is the classic example that most are familiar with. Your company has bench stock bins of nuts, bolts, and washers of various sizes in the maintenance shop, which are available as needed. Once per week, the hardware vendor would arrive at the shop and clean up the inventory by relocating any items that are not in their proper bins. The vendor would make a count of the items present, reconcile that with what was negotiated to have on hand, and fill the bins as appropriate. An invoice would be presented to the storeroom. Your company owns the stock; the vendor manages it.

Vendor managed inventory items will be negotiated by procurement or purchasing, and the Stores Stock Committee will be informed of the items that are controlled under VMI and be consulted when VMI items are added to or reduced.

VMI consideration will be given to any item received, tracked, and controlled by the storeroom. This includes parts, materials, and consumables related to plant maintenance and operations—any items that merit the approach given the gains to be made both in terms of financial and in gains made through service provided by others, balanced carefully against the impact to production.

VMI benefits the vendor by having contract agreements to provide a product and a service, guaranteeing revenue. VMI benefits your company in that the vendor provides the labor and the logistics trail for ordering and receiving inventory, and your company reduces the need to create numerous purchase orders. VMI lowers the ordering cost for an item.

To properly oversee the execution of VMI, the following is mandatory:

- Communication of what stock is managed by VMI
- List of which vendors manage what stock items
- Service schedule
- An invoice must be issued by the vendor and signed by the storeroom with every service/fulfillment visit
- Each vendor must have emergency contact information on file

- Specifications will be documented (e.g., the different grades of bolts, grades of stainless-steel fasteners, and seamless pipe) when required

After each service/fulfillment call, the vendor will report how it found the inventory and how it left the inventory.

I had an interesting childhood introduction to the idea of VMI, even though it wasn't called VMI back then. As a 16-year-old, I worked as a bagboy in a local grocery store. My duties included sacking groceries and carrying them out to the customer's car, sweeping/mopping, stocking shelves, and updating prices. From time to time, I would be on duty when some of the suppliers would show up to *service* their products. On Saturdays, the Coca-Cola distributor came in to work on the rows of Coca-Cola. He removed the product that was past its sell-by date and dusted off and moved the remaining product forward, filling the back sections with new product that he wheeled in on his dolly. Before he left, he removed a row of Pepsi from the shelf and replaced it with another row of Coke. He put the Pepsi product on the floor. About 30 minutes later, the Pepsi guy showed up and went through the same cleaning, facing, and stocking routine. Before he left, he replaced the ill-gained row of Coke with the original Pepsi, removed a row of Coke, and put Pepsi in its place. The displaced row of Coke was put unceremoniously on the floor. If the Pepsi guy ever went there first, the roles would have been reversed.

Apparently, this had been going on for years, if not decades. These two servicemen were good friends. This was just their way of saying hello to one another.

I asked the grocery store manager about it one day because I was the one who had to pick the surplus items off the floor and deal with them. The manager, George, told me about their decades-long ribbing and shared with me something I've never forgotten. George said, "I never have to worry about running out of Coke or Pepsi." We never did run out of Coke or Pepsi.

"Consignment" refers to inventory that is on the shelf in the storeroom or is on the shelf, off-site at a vendor's location. For on-site storage (your location), the upkeep of the component is the responsibility of your company. *Note:* You must follow the supplier's care instructions. For off-site storage (the supplier's location), the responsibility is on the vendor. Consignment stock typically consists of medium to high costs and is often considered to be items that are very consequential to production.

Examples would include:

- Medium to large motors
- Medium to large drives
- Medium to large gearboxes

With the Consignment process, the vendor owns the stock, and your company pays for the item when the item is used. Items identified for consignment will have a part number or a catalog number.

A typical scenario regarding consignment is where your company has the strategic need to have access to a 150-hp motor but not the room in the inventory budget. The vendor agrees to make the motor available when needed, with, or maybe without, a small fee (this should be negotiated). The motor can physically be located on-site, or it can be off-site at the vendor's location. When the motor is pulled from stores, a purchase requisition is issued, and an invoice for the motor is issued at that time.

With on-site consignment items, the vendor will specify care instructions, and they will need to be followed to the exact requirement. For off-site consignment items, procurement or purchasing will audit the items quarterly to ensure they are present and in good order.

Consignment items will be negotiated by procurement or purchasing, and the Stores Stock Committee will be informed of the items controlled under consignment and consulted with when consignment items are added to or reduced.

As with VMI consideration, Consignment consideration will be given to any item received, tracked, and controlled by the storeroom. This includes parts, materials, and consumables related to plant maintenance and operations—any items that merit the approach given the gains to be made both in terms of financial gains and in gains made through availability guarantee negotiated with others, balanced carefully against the impact to production. Having a critical spare set up as a consignment item would be foolhardy. That is my opinion. But I am unable to back my opinion up with facts or experience.

Consignment benefits the vendor in that the vendor has a contract agreement to provide a product and often collects a holding fee, therefore guaranteed revenue. Consignment benefits your company in that the vendor is making the item available at no, or very low, cost to the company.

To properly oversee the execution of consignment, the following are mandatory:

- A work order is required before removing a consignment item from stock.
- A purchase requisition must be made to account for the use of the item.
- There must be communication of what stock is managed by consignment.
- There must be a list of what actions to take for proper upkeep (Preservation Program).
- Notification to the vendor must be made immediately upon the use of any consignment item.
- An invoice must be issued by the vendor and signed by the storeroom with every use.
- Each vendor must have emergency contact information on file.
- Specifications (size, type, attributes, and details) will be documented when required.

After each use of a consignment item, the issue will be documented with a copy of the work order, the purchase requisition, and the invoice sent by the vendor.

I want to share a short story about the horrors that could become your reality for consignment items that are held at the vendor's location. A vendor will suggest holding an item for your use at its location for one of two reasons, or both:

- The vendor can better guard the reliability of the component in its care.
- The vendor has also promised it to another customer.

That second bullet can get you.

Years ago, I was visiting with a potential client in the Northeast. This client was a big player in the steel industry. The owner was a very successful businessman and was very well connected on a global scale. I really wanted to have him as a client.

Although he personally was a very busy man, wheeling and dealing, he was always generous with his time and truly had an open-door policy with me whenever I was in town.

I happened to be in his area one day on business with another client and asked if he had time for a quick visit. "Of course," he said, "I've always got time for you."

He was eager to show me his new business venture. When I stopped by, he took me out to his warehouse, and sitting in the middle of the floor was probably the biggest motor I had ever seen up to that point. I can't remember what size it was, but it might have been a 1,500-hp motor.

I asked him what the motor was for, and he said he had bought it as an investment. He said that he had it on consignment to a client and arranged to keep it in his shop under his supervision. He sold the idea by explaining that his team was better able to care for it. And as he told his client, "You don't have room for it, and it would take a week to get the old one out anyway. We can have this to you in 48 hours."

That all made sense to me, but I asked if he wasn't losing money on the deal by having to buy this motor, to sit and take up floor space. "No," he told me, "I'm charging them $1,000 a month to have it ready if they need it."

This had to be a $150,000 motor! "That's a long ROI," I said. My potential client told me that the ROI was shorter because he had also promised it and had it on consignment with nine other customers. I asked if they were all aware that he had a deal set up with other companies, and he said no.

"What if two people need it at the same time?" I asked him. He had gotten them to agree that if they needed it and he didn't have it, he would be required to pay a $10,000 penalty.

If you have followed the math, he made $10,000 in holding fees every month and paid for this huge motor in 15 months. After that, he made $10,000 monthly in holding fees as pure profit. This was a great investment scheme for him.

This is just a word to the wise.

I don't want to end the discussion on Consignment on a negative note. This is a creative method to have inventory readily available without incurring the cost to purchase until you need it or pay for years of carrying costs. Ramesh Gulati and Terry O'Hanlon provide a concise benefit to consignment in the book *10 Rights of Asset Management*: "Holding all critical parts in inventory can result in very high MRO storage expenses. Consider partnering with local industry and vendors for sharing some critical spares. Eliminating idle inventory is possible by negotiating delivery controls and establishing vendor trust" (p. 148).

SUMMARY

I always thought that the elements in this final Advanced-Efficient zone were an interesting collection of items. If I were to sit down and draft out some actions that I thought would make an effective organization more efficient, I think I would list problem-solving, root cause analysis and budgeting. Getting very good at these would make any job, any organization, and any effort more efficient.

The clean connections made at the end make this zone even more useful for rounding out the Storeroom Roadmap and tying up the efficiency effort.

Aside from RCA, which is a bit of an outlier, the process of Delivery is directly linked to the process of Kitting over in the Foundation-Efficient zone. The Stores Stock Committee, also in that earlier zone, directly links to the Monthly Budget, which is connected to Stores Issues vs. Budget. This makes tremendous sense if you remember that the finance department has permanent representation on the SSC. I think this is a real quote: "He who has the money makes the rules."

Finally, the tie-in between the very important process of Parts Standardization in the Intermediate-Efficient zone and Spare Parts Loan can be fruitful for any organization, including yours. As explained, an organization that insists on standardizing parts at a corporate level can capitalize on the idea of strategic spares. This is very high-level.

Of course, the Spare Parts Loan is further connected to VMI/ Consignment. This connection allows companies and industries to benefit from working with the big global and national parts suppliers. This is all part of the bigger strategic sourcing effort.

WORKBOOK

The workbook for Chapter 7 is the final accounting of how you feel your organization is navigating the elements of the Storeroom Roadmap. If you've answered all the questions I've asked over the chapters and pages of this book, you have a real gut feel about what needs to be done to get your storeroom on the road.

Here is the assignment for Chapter 7. Simply fill in Table 7-7 with a Y for yes or an N for no.

TABLE 7-7 Answer Y or N

	Y or N	N/A
Do you have formal process guides created for the following?		
Root Cause Analysis and/or problem-solving		
Delivery of kitted parts		
Spare Parts Loan		
VMI/Consignment		
Does your storeroom play an active role in controlling the maintenance budget?		
Does the storeroom report on items removed from stores that are very expensive items?		

If you haven't already thought about showing these short videos to your Stores Stock Committee at its monthly meetings, might I suggest you consider doing that? Table 7-8 has the links to the final videos.

TABLE 7-8 Topic Links

Title	Maintenance Innovators, Inc., Website Link
Spare Parts Loan	https://maintenanceinnovators.com/aiovg_videos/spare-parts-loan/
VMI/Consignment	https://maintenanceinnovators.com/aiovg_videos/vmi-consignment/
	YouTube Dr. John's Maintenance Minute Link
Spare Parts Loan	https://www.youtube.com/watch?v=mLEDIWRSPY8
VMI/Consignment	https://www.youtube.com/watch?v=8GfXEt6teWM&t=21s

CHAPTER 8

THE ROAD BEYOND THE MAP

Roads? Where we're going, we don't need roads.

—Doc Brown, *Back to the Future*

There's a small town in my home state of Oklahoma, up near the panhandle, called Slapout (pop. 4). I've never been there myself, but I understand that the town was built where you ran *slap out* of road. This might be an urban (or rural) legend, but those who have been to Slapout can attest to the fact that there isn't much there. The road less traveled, so to speak.

Maybe Doc Brown got it right when he told Marty and Jennifer that roads weren't needed in the future. I guess there will be no place for the Slapouts of the future. No road, no town. If there are no roads, will there still be maps? Does the road go beyond where the map ends?

Have you ever thought about a time before we had charts or maps or other navigational devices? How did explorers navigate? What comforted them with the calming knowledge that they were going in the right direction? What did Leif Erikson, Christopher Columbus, and Charles Lindbergh use to navigate? The stars. Of those three, only Lindbergh was certain there was something beyond the vastness.

There is a very famous map painted on the wall of the Vatican in the Hall of Maps. In fact, all the maps in that hall are famous. The very fact that these marvelous works of art and science even exist is amazing. Around 1580, Pope Gregory XIII commissioned famed mathematician, astronomer, and cosmographer Ignazio Danti to document the vast extent of his [the pope's] worldly dominion. Apparently, it was quite popular for the aristocrats of that era to flaunt their wealth and status by commissioning great works of art, including maps. The Vatican has an entire wing covered in the most elaborate maps you will ever see. This is the Hall of Maps. But it's one map in particular that I'm fascinated with.

Curiously, one of the more spectacular maps has the heroic image of Christopher Columbus. I had heard about this map and that there was a painting of Columbus on it. I was not ready for the grandeur of the map when I saw it in person. Figures 8-1 and 8-2 are the pictures I took of this map. Believe me, the black-and-white images you are seeing do not do this painting justice.

FIGURE 8-1 Hall of Maps.

FIGURE 8-2 Christopher Columbus.

Through Eternity Tours has perhaps the most moving description of the stark image of Columbus:

> Columbus, dressed in the triumphant guise of a Roman emperor as he speeds away from Genoa and down the Ligurian coast. Compass grasped in his hand like his very own sceptre of power, he sits forward in his golden chariot as his white toga is whipped around his shoulders by a sea breeze. No less a figure than the sea-God Neptune himself guides its course, standing on a massive sea-shell lashed to the backs of four sea-horses thrashing the foamy sea with their hooves. A merman trumpets through a conch as he leads the way, carrying aloft a banner confirming the venerably bearded man's identity: "Christopher Columbus of Liguria, Discoverer of the New World." (*Througheternity Blog*)

I offer this unusual chapter start to get to the last sentence you just read. "Christopher Columbus . . . Discoverer of the New World." For those of you who have made it to this point with me in our journey through the Storeroom Roadmap, this is where I leave you, and you must go forward on your own. I've taken you as far as I can on the paved road. What lies ahead is completely up to you. You are the Discoverer of the New World for your organization.

Because I can't let this go, I want to offer just a bit more information for any history buffs reading this book. We have the same Pope Gregory XIII to thank for the Gregorian calendar we use and leap year as we know it. This was a pope who got things done! According to Associated Press journalist Leanne Italie as reported in the *Tulsa World* newspaper:

> Enter Pope Gregory XIII, who calibrated [the length of a year] further. His Gregorian calendar took effect in the late 16th century. It remains in use today, reducing drift [the errors in timekeeping] to mere seconds.
>
> Why did he step in? Well, Easter. It was coming later in the year over time, and the pope wanted Easter to remain in the spring.
>
> He eliminated some extra days accumulated on the Julian calendar and tweaked the rules on leap day. It's Pope Gregory and his advisers who came up with the gnarly math on when there should be or shouldn't be a leap year. (p. D5)

Like Pope Gregory, who worked to improve the calendar and even how we recognize the years, ours will also be a lifetime of continuous improvement.

CONTINUOUS IMPROVEMENT

Figure 8-3 was first shown in *The Reliability Excellence Workbook: From Ideas to Action* to demonstrate the change model used to illustrate where continuous improvement fits within culture change.

The journey you and I have been on using the Storeroom Roadmap has taken us from the *status quo* to a new *culture*. You established early in the journey where you were. You took Earl Porter's message to heart: "Even if you know where you're going, you're still lost if you don't know where you are." That message is a testament to the value of measuring and understanding the status quo. The first element of the change model is to confirm the status quo. Like any trip preparation, you must have a checklist. You can tick "status quo" off the list.

From status quo in the figure, the leader leads the organization toward the vision. If you've followed any advice, completed any homework, or watched any of Dr. John's Maintenance Minute videos, you can't help but see the future. You, the leader, are also responsible for helping others see the future. I'm sure Christopher Columbus had to remind his crew that land was somewhere, just over the horizon. This is the vision. You determine outcomes by influencing the future. I think Tom Moriarty would agree.

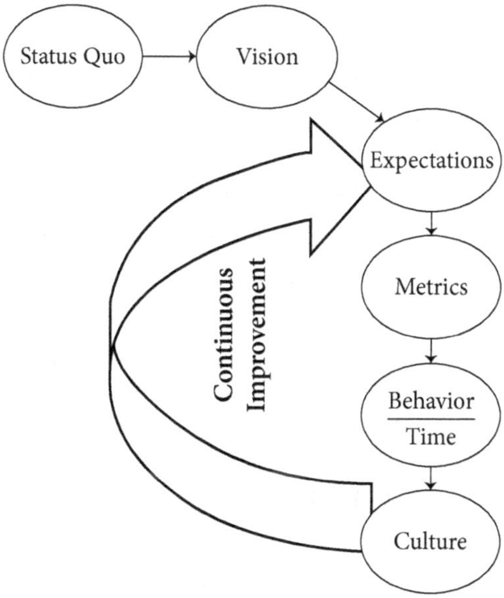

FIGURE 8-3 Change model with continuous improvement.

Few could argue that by enacting many of the best practices in this book, you are, in fact, setting newer and higher expectations for your storeroom associates. And in my opinion, you're setting them pretty darn high. This book was sprinkled with advice and wisdom from many of the most respected sources in the industry. When was it ever a bad idea to hold our people to a higher standard? I think it would be insulting to doubt their ability to rise to that level. I've heard this quote attributed to Michelangelo: "The greater danger for most of us lies not in setting our aim too high and falling short; but in setting our aim too low, and achieving our mark."

Your expectations were established when you first subscribed to the four rules that started this entire voyage. And continued through SMRP-Gulati-Slater-Palmer, et al.

Metrics? KPIs? If you've taken notes and look back on what we've covered, you will agree that you have been provided with enough best-in-class target values to set your course and perform a gap analysis between where you are and where you want to be. Life is about closing those gaps. Don't forget the Big 3, the Sergeant Majors.

If you look at the change model in Figure 8-3 one more time, you'll notice that metrics drive behavior. And a different behavior over time creates a new culture. It turns out that culture is simply a manifestation of what is allowed to happen over time. Go back to Moriarty. You can influence the future.

In my class, when I use this change model, it is here at the culture change point that I draw the continuous improvement arrow. An organization should wait until a critical mass is formed at the new cultural point before it moves en masse. Notice that the tip of the continuous improvement arrow is pointed at expectations. Continuous improvement should move the organization to newer, and presumably higher, expectations.

At the end of this printed roadmap, my challenge for you is to make that sweeping continuous improvement drive to a higher expectation for your storeroom and your organization.

KIDS TODAY

My business leans about 50/50 between maintenance consulting and teaching, on the one hand, and storeroom consulting and teaching, on the other. In either engagement, I've noticed that the kids today expect to use some type of technology as a tool to complete their routine work. If a young, new employee signs on to work in your storeroom and on the first day, you hand the employee a clipboard and a pencil, you are likely to get a blank stare coming back at you. "Where's my tablet?" "Where's my drone?" "Where is my digital interface?" This is the 21st century. Where's my flying car?

I realize those of you reading this book could be doing so years from its first publication. But to give you a scale of our technology gap, I'd like to list a few experiences I've had in the last year.

Last week, I was in a storeroom where the associates were gushing over bar code capability that was coming to them sometime *this* year. They spoke about it like it was Gutenberg's movable type. "Boy, big things are going to happen for us now!" Bar coding.

Within the last year, I led a project to move a storeroom from outside (yes, it was outside, in the dirt, in the United States) to a "new" building.

My associates and I had to dig almost every component out of the dirt and clean it before moving it. Most of the items were in dilapidated crates that had been three-quarters stuck in the mud. I wouldn't have bothered keeping any of that stock. It was all ruined. "But John," I was told, "this is $2 million worth of inventory."

Late last year, I was in a storeroom that used augmented reality and the cloud.

The storeroom industry is ripe for a revolution, and we are going to see a wave of technology like you've never seen before.

I am in a partnership that has created and oversees several certification programs. In November of each year, we have the rare opportunity to merge two programs. During that month, each year, the reliability engineering candidates attend in person at a technology center for a week of hands-on technology application and study, while at the same time, we are always conducting an in-person storeroom certification class. On Thursday of that week, it is 3-D printing day. The hands-on lab is meant for the reliability engineers, but we have the storeroom class sit in for an hour or so of the demonstration of 3-D printing. Why? Because we believe that this technology is going to play a big role in the storerooms of the future.

It might be that your storeroom has a 3-D printer and you are using it. I think that design engineers and OEMs will start colluding with parts that can be printed on a 3-D printer. Can you imagine the advantage that a design company would have in selling you on a machine for which it could email you the spare part? My goodness. Somebody ought to write that down.

The instructor who teaches the 3-D printing lab for me and my company told the class last November that NASA had sent the astronauts on the space station the code to print a wrench that they needed. According to NASA.gov, "This isn't the first 3D-printed object made in space, but it is the first created to meet the needs of an astronaut. When International Space Station Commander Barry Wilmore needed a wrench, NASA knew what to do. They 'e-mailed' him one. This is the first time an object has been designed on Earth and transmitted to space for manufacture." Figure 8-4 is an image of that wrench.

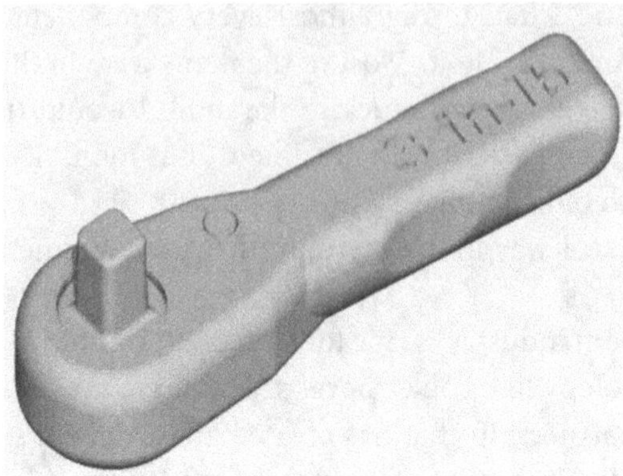

FIGURE 8-4 3-D wrench made in space (image from the NASA website).

I believe that machines in the future will be designed and built with material that allows for the replacement parts to be printed. Can you imagine the implications?

What limits this application is the material that is needed to print with. Someone, somewhere, is working on printing with the exact material you need. I wouldn't be surprised if this were a black government program right now. My old military mind is thinking of the national security benefit of printing capabilities on the front line to execute the mission. Need a shovel to dig a foxhole? Fire up the printer—I'm sending you the program.

OLD NEWS

We've seen a lot of this before. I tell folks we are about 100 years from having those *Star Trek* replicators. You know the ones. Walk up to the vending machine–looking "thing" and request "Earl Grey. Hot." With technology invading every facet of our lives now, even instant gratification isn't soon enough.

Try this and see what happens. The next time you are with your friends, discuss wanting to take a cruise with your family. Keep that conversation going for a few minutes. You will instantly receive advertisements for cruise packages on social media. Everyone reading the previous sentence has had that experience and knows what I'm talking about. The speed of business demands a different paradigm. Good things no longer come to those who wait. Waiting for good things is a good way to get run over.

I said this was old news. Everything I've written about has already happened. It's no longer new. But it can be improved.

Figure 8-5 first appeared in *The Reliability Excellence Workbook: From Ideas to Action*. I still have the actual page from *Fortune* magazine where I got this image. I carried it around in one of those plastic document protectors for years. I used to use it as an ice-breaker for my classes.

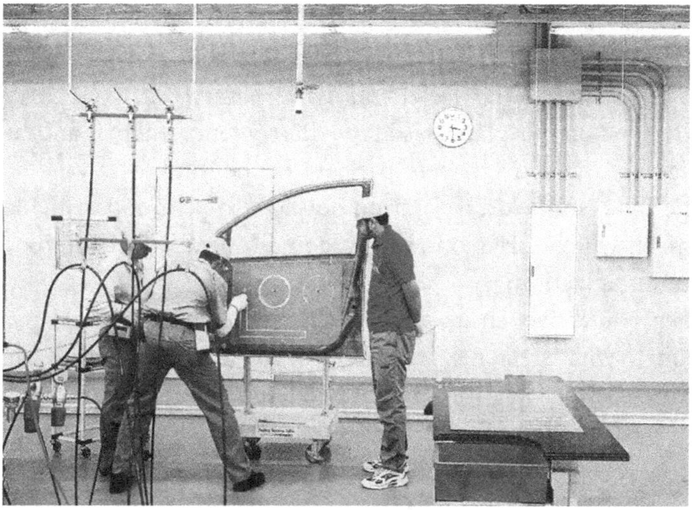

FIGURE 8-5 Nissan technicians.

The article that accompanied this photo in October 2007's edition of *Fortune* magazine indicated that Nissan technicians were exploring ways to put a door together using as few tools as possible (Taylor, p. 130).

In 2007! That's decidedly *not new*!

The next time you complain about maintenance technicians not writing down the parts that they are removing from the storeroom, please think about how they can print tools in space and build cars without tooling. Your problems sound ridiculous when you think about how the world has advanced. As I've said, "If you can't master pencil and paper, technology isn't going to help you."

THE STOREROOM OF THE FUTURE

I don't know if the storeroom of the future will have 32 processes and practices. There might not even be a storeroom in the future. The idea of investing company capital in a bunch of spare parts that you may or may not ever use (before they deteriorate) doesn't make sense in this technologically advanced world. One thing is for certain—technology will continue to advance. The world demands it. Instant gratification? Who has the time?

Give some real thought to why you have spare parts in the first place. You might not remember this sentence from John Moubray's quote in Chapter 5: "If it could be done instantly, there would be no need to stock any spares at all."

"If it could be done instantly . . ." and we are getting closer to that every day *". . . there would be no need to stock any spares at all."*

Right now, on this planet, as I type this sentence, there is another company digging a destroyed spare part out of the mud to salvage it, and at the same time, some kid is figuring out how to print a part made out of titanium. In which direction are you going to lead your storeroom team and your organization?

When you arrive, en masse, at a changed culture in your spare parts operation, to what greater expectation will you take your spare parts effort? Do you determine outcomes by influencing the future, and are you the Discoverer of the New World?

References

Advanced Storeroom World-Class Principles. CSM/CST, American Reliability & Maintenance Institute, PowerPoint course presentation.

Arena Best Practices. "What Is the Right Part Numbering Scheme for You? Intelligent vs. Non-Intelligent," www.arenasolutions.com/resources/articles/part-numbering/.

Blache, K. M. (2023, October). "Use Best Practices for MRO Inventory," *Efficient Plant*, p. 6.

"Christopher Columbus Hitches a Ride in the Vatican Museums' Hall of Maps," *Througheternity Blog*, www.througheternity.com/en/blog/art/columbus-hall-maps-vatican-museums.html.

Gulati, R. (2013). *Maintenance and Reliability Best Practices*, 2nd ed. New York: Industrial Press.

Gulati, R., & O'Hanlon, T. (2021). *10 Rights of Asset Management: Achieve Reliability, Asset Performance and Operational Excellence.* Ft. Myers, FL: Reliability Web.

Ho, M., & Hodkiewicz, M. (2013). *Glossary of Reliability Terms and Definitions for Maintenance of In-Service Assets*, 2nd ed. University of Western Australia and CRC Mining, Australia.

Ingalls, P. (2019). *Making Friction Fiction: The Art & Science of Greasing.* TBR Strategies.

International Standard, ISO 14224, Second Edition 2006-12-15. Published in Switzerland.

International Standard, ISO Standard 55002:2014. "Asset management: Management systems, 1st ed. Published in Switzerland.

Italie, L. (2024, February 28). "The Power of Leap Year," *The Tulsa World*, p. D5.

Kellett, K. *Blog*. Travelrat's Travels, www.travelrat.wordpress.com.

Kiyosaki, R. T. (1998). *Rich Dad Poor Dad: What the Rich Teach Their Kids About Money—That the Poor and Middle Class Do Not!* New York: Warner Business Books.

Lee, R. D. (2018). *The Maintenance Insanity Cure: Practical Solutions to Improve Maintenance Work*. South Norwalk, CT: Industrial Press.

Liker, J. K. (2004). *The Toyota Way: 14 Management Principles from the World's Greatest Manufacturer*. New York: McGraw-Hill.

Maintenance Innovators, Inc. Dr. John L. Ross, Jr. Website, established 2007, www.maintenanceinnovators.com.

Maintenance, Repair & Operations (MRO). TBR Strategies, PowerPoint course presentation.

Maintenance, Repair and Operations (MRO) Best Practices. (2017). Survey Results. Ft. Myers, FL: Reliability Web.

Materials Management for Maintenance, 8-Hour Overview. Marshall Institute, V5.0. Raleigh, NC, PowerPoint course presentation.

Moore, R. (2004). *Making Common Sense Common Practice: Models for Manufacturing Excellence*. Linacre House, Jordan Hill, Oxford, UK: Elsevier.

Moriarty, T. J. (2020). *The Productive Leadership System*. South Norwalk, CT: Industrial Press.

Moubray, J. (1997). *RCM II: Reliability Centered Maintenance*, 2nd ed. South Norwalk, CT: Industrial Press.

MRM 1, Maintenance and Reliability Management. North Carolina State University, Raleigh, NC, Diploma Series,

MSC Materials Management for Maintenance. Marshall Institute, V7.0.1. Raleigh, NC, PowerPoint course presentation.

OSHA, 1915.82(a) (2). For landside areas. Table F-1.

Palmer, R. (2019). *Maintenance Planning and Scheduling Handbook*, 4th ed. New York: McGraw-Hill.

Pellegrino, A. (2023). *Reliability Workshop for Executives*. Spartakus Technologies, PowerPoint course presentation.

Pope, A. (1773–1774). *An Essay on Man: Epistle 1.*

Regan, N. (2012). *The RCM Solution: A Practical Guide to Starting and Maintaining a Successful RCM Program*. New York: Industrial Press.

Ross, J. (2018). *The Reliability Excellence Workbook: From Ideas to Action*. South Norwalk, CT: Industrial Press.

Ross, J. (2019). *Cover Your A$$ets: Asset Management at Your Place and at Your Pace*. South Norwalk, CT: Industrial Press.

Ross, J. (2024). *Best Practices in Maintenance and Reliability*. Maintenance Innovators, Inc., PowerPoint course presentation.

Schillaci, W. C. (2018). "Is Your Worksite Lighting Meeting OSHA Standards?" *EHS Daily Advisor*, https://ehsdailyadvisor.blr.com/2018/08/worksite-lighting-meeting-osha-standards/.

Slater, P. (2017). *Spare Parts Inventory Management: A Complete Guide to Sparesology*. South Norwalk, CT: Industrial Press.

SMRP Best Practices, 5th ed. (2017). Society for Maintenance & Reliability Professionals.

SMRP Guide to the Maintenance and Reliability Body of Knowledge (2009). Society for Maintenance & Reliability Professionals.

Stanley, T. (2024, January 7). *Tulsa World* started a series called "Faces of Hope." *The Tulsa World*, p. A1.

Sutherland, J. (2014). *Scrum: The Art of Doing Twice the Work in Half the Time*. New York: Crown Business Books.

Suzuki, T. (Ed.). (1994). *TPM in Process Industries*. New York: Productivity Press.

Taylor, Alex. (2007, October 29). "Nissan's Radical Chic," *Fortune* magazine, p. 130.

Towles, A. (2019). *A Gentleman in Moscow*. New York: Penguin Random House.

TPM/TPR Coordinators Workshop. Marshall Institute, V4.5P. Raleigh, NC, PowerPoint course presentation.

Wiseman, L. (2017). *Multipliers: How the Best Leaders Make Everyone Smarter*. New York: Harper Business.

"Wrench." (2015). NASA 3D Resources, https://nasa3d.arc.nasa.gov/detail/wrench-mis.

Wright, N. C. (2017). *The Death of Reliability: Is It Too Late to Resurrect the Last, True Competitive Advantage?* South Norwalk, CT: Industrial Press.

Index

Page numbers followed by *f* and *t* refer to figures and tables, respectively.

www.ingramcontent.com/pod-product-compliance
Lightning Source LLC
LaVergne TN
LVHW052254200926
848481LV00002B/365
* 9 7 8 0 8 3 1 1 3 6 9 4 9 *